Brian Weinstein
Nancy Weingate
Lisa Byrd
Prt 451 9-16-96
@TN/SLC

Availability Engineering and Management for Manufacturing Plant Performance

PRENTICE HALL INTERNATIONAL SERIES
IN INDUSTRIAL AND SYSTEMS ENGINEERING

W. J. Fabrycky and J. H. Mize, Editors

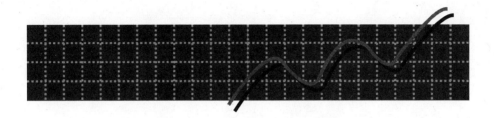

Availability Engineering and Management for Manufacturing Plant Performance

Richard G. Lamb, PE, CPA

Prentice Hall P T R
Englewood Cliffs, NJ 07632

Library of Congress Cataloging-in-Publication Data

Lamb, Richard G.
Availability engineering and management for manufacturing plant
performance / Richard G. Lamb.
      p.   cm.
Includes bibliographical references and index.
ISBN 0-13-324112-2
1. Factory management. 2. Plant maintenance.   I. Title
TS155.L248 1995
658.2 -- dc 20

94-49535
CIP

Acquisitions editor: Bernard Goodwin
Editorial assistant: Diane Spina
Cover design: Design Source
Cover design director: Jerry Votta
Interior design: Gail Cocker-Bogusz
Production coordinator (Buyer): Alexis R. Heydt

Figure 13-6 is from Jay R. Galbraith, *Designing Complex Organizations* (pg. 15), © 1973 by Addison-Wesley Publishing Company, Inc. Reprinted by permission of the publisher. Text on page 140 is copyright © 1984. Electric Power Research Institute. EPRI NP-3364. *Commercial Aviation Experience of Value to the Nuclear Industry.* Reprinted with permission. Text on page 161 is copyright © 1989. Electric Power Research Institute. EPRI GS-6266. *Demonstration of an Availability Optimization Method.* Reprinted with permission. Text on page 171 is copyright © 1981. Electric Power Research Institute. EPRI NP-1567. *Human Factors Review of Power Plant Maintainability.* Reprinted with permission.

The publisher offers discounts on this book when ordered in bulk quantities. For more information, contact Corporate Sales Department, Prentice Hall PTR, 113 Sylvan Avenue, Englewood Cliffs, NJ 07632. Phone: 800-382-3419 or 201-592-2498; FAX: 201-592-2249. E-mail: dan_rush@prenhall.com.

Printed in the United States of America

10 9 8 7 6 5 4 3 2 1

ISBN 0-13-324112-2

Prentice-Hall International (UK) Limited, *London*
Prentice-Hall of Australia Pty. Limited, *Sydney*
Prentice-Hall Canada Inc., *Toronto*
Prentice-Hall Hispanoamericana, S.A., *Mexico*
Prentice-Hall of India Private Limited, *New Delhi*
Prentice-Hall of Japan, Inc., *Tokyo*
Simon & Schuster Asia Pte. Ltd., *Singapore*
Editora Prentice-Hall do Brasil, Ltda., *Rio de Janeiro*

To JoEllen,
my lovely young bride of many years,
and our wonderful children,
Danna and Joseph
whose generation will surely advance
the knowledge and writings of ours.

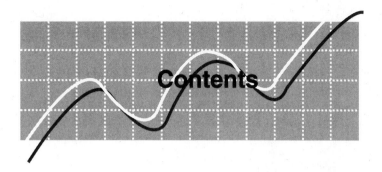

Contents

Part 2   The Conceptual Design Phase

Chapter 3   Availability Engineering and Management and the Conceptual Design Phase   38

Chapter 4   Preparation for Project Execution   59

Part 3   Basic Design Phase

Chapter 5   Foundations to Maintain Integrity of the Availability Scheme   71

Part 4   The Detailed Design Phase

Chapter 11   Development of Maintenance Tasks
and Procedures   182

Chapter 12   Determination of Resource Levels for
Operational Availability   209

Chapter 13   Organization Design for Maintenance
Operation Functions and Availability Management   232

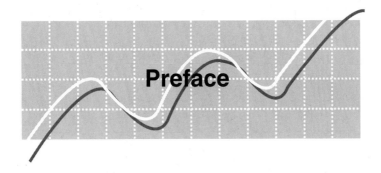

Preface

Purpose of the Book

A manufacturer's overall business success will be determined by the following integrated dimensions of performance:

- Fitness of the manufacturer's products for use and their producibility.
- The degree to which its manufacturing plant as a system of equipment is physically available to perform.
- The degree degree to which the manufacturing process and its operation can efficiently produce the products.
- Maximum effectiveness and efficiency of sales and distribution of the resulting productive capacity.

These primary money-making subsystems almost completely determine the manufacturer's competitiveness, profits, and productivity of working and capital assets. They are identified here as a point of reference to define the purpose of this book, which is to describe the principles and the approach necessary to develop and manage the plant's mechanical availability to perform.

Pretty Limited !

The discipline to serve this purpose is little known outside its long and successful applications in the defense and aerospace industries. From there, it has evolved to some degree in the nuclear power industry, which has adopted some of its techniques and management practices.

Until now, the field has not been reinvented and advanced to serve manufacturing enterprises and their plants. The available literature has focused on the product (i.e., weapon system) rather than the manufacturing plant. This book explores for the first time a comprehensive solution design framework. It is also the first time the field has been integrated with widely known principles of plant startup and production system commercial operations.

Therefore, the purpose of the book is important. History has left the development and management of availability performance with great room for advancement. Thus, the book is the key to the manufacturer's currently most potent investment opportunity. In fact, this opportunity is one that will most decide each manufacturer's continuing competitiveness.

The book's publication is also timely. Many manufacturers have already expressed their desire for availability performance. It is common to find that pivotal corporate, operating company, and plant-level policies have already been established. They may directly mandate that reliability, maintainability, and, therefore, availability are challenges to be solved. This is the same for existing facilities and future capital projects. There are also many indirect mandates for availability engineering and management. This is because goals such as asset utilization, perfection in manufacturing, etc., are dependent upon the availability discipline for defining their role in overall business success.

Availability Engineering versus Total Productive Maintenance

Maintenance is the largest activity area within availability performance as a core business process. In other words, the "production" dimension or major "product" of availability design and management is maintenance. Therefore, the purpose of maintenance operations (including maintenance activities done by operating personnel) is to deliver the currently required availability performance the plant, as a system of equipment, is capable of providing.

The subject of the book, however, is not total productive maintenance. Nor is it total productive maintenance with a different name or variation in approach. Availability engineering and management has a much larger domain. It subsumes the issues presented in total productive maintenance literature.

The primary goal of total productive maintenance is to achieve maximum equipment effectiveness. The goal of availability engineering and management is to determine and achieve the availability performance necessary to the manufacturer's corporate, operating company, and plant-level business performance and leadership. Therefore, total productive maintenance is driven to maximize the parts of the plant as a production system. By comparison, availability engineering and management works to suboptimize the parts, thus maximizing the whole.

The design processes of total productive maintenance and availability engineering and management are very different. The processes used in total productive maintenance are problem solving approaches like those used in focused quality improvement projects. Those used in availability engineering and management focus on designing and managing a production business system.

Not a totally true statement

This difference highlights the distinction made by business process reengineering literature between quality improvement and step level advancements in business results.

The point of making this distinction is that the subject of this book is a critical advancement. It has long been searched for by management professionals. Manufacturers are now bombarded by methods such as total productive maintenance, reliability engineering, availability (RAM) modeling, life-cycle and activity-based costing, reliability centered maintenance, etc. However, the immense potential of these is not achievable outside the system of processes to design and manage dynamic, cost-effective plant level availability performance. In other words, the field of availability forms a basis for these methods. Furthermore, without availability engineering these methods can easily reduce rather than advance plant performance.

Readers and Their Needs

This book is written for technical and management professionals in the continuous production industries. This includes industries such as oil and gas production, pipelines, petrochemical and chemical, refining, pulp and paper, power, food, etc. It also includes industries that produce discrete products. The common thread is that they require substantial systems of equipment. However, it could easily apply to noncontinuous production operations (i.e., job shops) in which equipment and its maintenance is critical to productive capacity.

The book is written for people from many organizational levels and functions. Design and management of availability performance spans the domains of technical and management roles in general business and production operations management. This integration begins with the initial phases of design and continues through the plant's operating life.

Therefore, this book addresses the concerns of the following readers:

Plant and general managers. These readers need to understand the challenges, business consequences, and approaches to achieving maximally dynamic and cost-effective availability performance.

Managers of plant capital projects and performance advancement programs. They must be able to determine whether the planned approach is appropriate and comprehensive.

Plant design engineers and business discipline specialists involved in executing the planned approach. These readers have a dual need: To incorporate the requirements of availability design in the traditional approach to capital projects and to effectively participate in the subject projects and programs.

Maintenance operation managers, personnel and specialized experts. These readers need a process to guide them as they align maintenance operations to be in the business of delivering the availability performance the plant is capable of.

Computer systems integrators. Availability performance requires the integration of computer systems and software. Thus, these readers need comprehensive detail to give them vision and guide them in this challenge.

Process safety and environmental management professionals. Their need is to understand a field that is fundamental to the mechanical integrity of critical equipment.

The book is important to another group of readers outside the continuous production industries. The field was first developed decades ago by the military for weapons and aerospace systems. Practitioners in those sectors are now seeking employment in industry as military spending decreases. However, they are not versed in the application of the field to the industries now discovering its importance. Furthermore, their experience in the field of availability engineering and managements is in the design of *products* rather than *manufacturing systems*. Nor are they familiar with the design and management activities for these plants.

The book is also important to the many management consultants in the field of business process reengineering. The book's subject is repeatedly revealed to be a primary target for reengineering. It is one of several that will most frequently be found to affect their clients' business results. Further-

more, the field of availability engineering and management is truly representative of the innovation reengineering practitioners seek.

What To, Why To, and How To

Everybody knows the saying, "To give a fish is to feed a person for a day, to teach how to fish is to feed a person for a lifetime." This philosophy is germane to this book.

The book is written to teach the availability discipline as a system of design and management activities. Thus, it is a *what to*, *how to*, and *why to* book. The description of *what to* and *how to* is used as a foundation to explain *why to*. Accordingly, it has the following structure:

- The discipline is presented to track the phases of development in the plant's life cycle. The phases begin with design and progress though the plant's producing life.
- A flow of top-level design and management processes are charted for each life cycle phase.
- Each process is then presented in a flowchart and explained as development tasks and management subprocesses.
- Throughout the book these three levels of work flow from the top down and are used as a platform to explain the *why to* of availability engineering and management.

Readers will benefit from the book's organization in two ways. First, they will have information structured in a manner that can be directly applied to plan and execute programs for achieving necessary availability performance. Second, they will gain the philosophical knowledge to think originally in these pursuits rather than be restricted to a recipe.

Presenting the *how to* of the many individual analytical tools and probability theory is potentially counterproductive. The literature, expertise and services for treating the *how to* of these items are already widely available. To rigorously explain them here is a potential distraction to establishing the reader's ability to fish.

However, knowledge of these tools and theories are fundamental to presenting the availability discipline. Therefore, the book will still explore their purpose, design issues, processes, and why and when they are necessary. To teach the reader to fish, it is still necessary to describe how these known pieces are applied and then become part of managing availability performance. This, as part of detailing the availability discipline, is the problem that remains to be solved. The *how to* of the pieces has already been solved and practiced.

Capital Projects and Existing Plants

As mentioned, the book presents availability engineering and management with respect to the plant's life cycle. The life-cycle phases include the initial feasibility and design phases, construction and startup, and the plant's producing life. Improvements, modifications, and capital projects will also be part of the producing phase.

This approach is suitable for the development of availability performance beginning with the plant as a capital project as opposed to projects for advancing the availability performance of an existing plant. The described development and management processes are equally relevant to both.

Compared to the capital project, reengineering availability performance in an existing facility is a greater challenge. The plant has already taken on a life of its own—albeit one of suboptimal business results. A natural response is to approach the advancement of availability performance with piecemeal techniques. Focusing on individual plant items, functions, and resources. Plant-level performance becomes a "blackbox" that these individual initiatives hope to affect.

The alternative is the process described in this book. It evaluates the elements of plant availability performance that have been developed by intent or default; determines an obtainable ideal availability scheme; and formulates a realistic master plan to reach that ideal over time.

In that sequence, the process of the book focuses on the analysis, development, and management of maintenance operations. This begins with the recognition that its business is to deliver the current necessary plant-level availability. Thus, management's goal is to align plant field and support functioning to achieve its desired performance. This alignment will complement the availability factors that are now generally inflexible.

Projects to expand or modify existing plants fall between these two extremes. Thus, they too present a challenge beyond that for a straight forward capital project. However, once again the process is still applicable.

The life cycle approach of this book will benefit all types of readers. It serves as a baseline of knowledge of availability engineering and maintenance. From there readers can devise an approach to their current challenge. In other words, an overriding goal of this book is to enable its reader to develop and implement a program for any plant—new or existing.

Acknowledgement

This book is the development of a long sought solution approach to a fundamental area of plant performance. In industry, it is the area that has been least charted and is now management's greatest opportunity to advance production plant performance.

The development of the solution process of the book required the contribution of many people—it is teamwork performed on a large informal scale. I have had the occasion to discuss the field with many people. In long and short discussions, they shared the learning and details of their experience. Thus, they tested, challenged and advanced our combined understanding of the evolving solution approach. When they read this book, they will recognize their influence. I am also personally blessed by the gift of their time.

There were several people who donated their time as reviewers to both "sanity check" and advance the book's content. They are Dr. Benjamin Ostrofsky of the University of Houston; Mr. Rudy McCamish of DOW Chemical, Texas; and Dr. Woodrow Roberts of DOW Chemical, Louisiana. The many discussions I had with these gentlemen were always uplifting and encouraging.

Dr. Ostrofsky—who has practiced and advanced the field since its inception decades ago in the aerospace industry—assured that the reinvention of the discipline for industry and the application-centered approach to presenting a vast, complex subject remained technically and theoretically correct. His experience in plant design was also important in that role. Mr. McCamish and Dr. Roberts—as thought leaders and practitioners of many years in plant engineering, operation, maintenance and reliability—assured that the field's presentation remained firmly connected to the reality of production plants.

And finally, all of us as champions of the book's subject are indebted to JoEllen Roper-Lamb. She provided the time as lay-reader to test whether the vast, complex subject had been presented effectively in word and diagram. By doing the word processing and helping with the art, she also allowed me to conserve energy for editing the many drafts necessary to communicate a difficult subject and the contribution of its many champions.

PART ONE

**Definition and Value of Availability
Performance**

CHAPTER 1

Definition and Goals
of Availability Engineering
and Management

Availability as a Core Subsystem
of Plant and Business Performance

It is easy to understand the importance of availability engineering and management to business success if the plant is viewed within a system of money-making subsystems. Availability performance is one of these subsystems. Its strategic significance is shown in Figure 1-1.

Business success, therefore, can be viewed as a function of the following four integrated money-making subsystems.

- Product composition, fitness for use, and producibility.
- The physical availability of the facility to perform its production process.
- The production process and its support activities.
- Sales and distribution of the productive capacity created by the previous three subsystems.

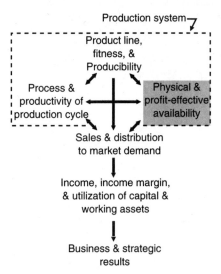

Production system

Product line,
fitness, &
Producibility

Process &
productivity of
production cycle

Physical &
profit-effective
availability

Sales & distribution
to market demand

Income, income margin,
& utilization of capital &
working assets

Business & strategic
results

Fig. 1–1 Availability performance as a
core element of business results.

The design of a production system begins with its products and their producibility. The design of the product and how the product will be produced are iterative activities because they are mutually constraining. How the product will be produced is the integration of the plant's availability and production process schemes. Product producibility, the availability to perform, and the production process jointly determine productive capacity and its economics. The degree to which the capacity is utilized will depend upon the owner's ability to sell and distribute the plant's products.

The four subsystems almost completely determine the owner's competitiveness, income, and productivity of working and capital assets. Therefore, how well the four subsystems are integrated by design and then dynamically managed is critical because short- and longer-term change are basic realities of the product, production process, and sales and distribution subsystems. The subsystem for physical availability must respond to the resulting operating conditions.

If any of these money-making subsystems are not appropriately designed and managed, the plant as a production system and the owner's business results will suffer.

Few would disagree that the availability subsystem has great room for advancement. This has implications for the goal of corporate, operating company, and plant management to identify the investment opportunities that will most increase their organization's business performance. Attention to the availability subsystem will be among the most attractive investments for the organization's energies and resources.

Attributes of Availability Performance

The exploration of availability engineering and management begins by defining availability performance and its nature. The key aspects to grasp, at this point, are as follows:

- Availability as a probability distribution function.
- Availability, reliability, and maintainability defined as performance characteristics.
- Reliability as the interface characteristic between the product and process subsystems.
- The three subtypes of availability performance.
- The top-level factors of reliability and maintainability.

Availability as a Probability Distribution

Availability performance is characterized as a probability distribution of the ratio or percentage of time the plant or its subsystems are available to perform. This ratio is represented in Figure 1-2, which shows performance as a mean with a range to both sides. This distribution is associated with planned operating conditions. If the plant is operated differently, the shape and position of the probability distribution will change.

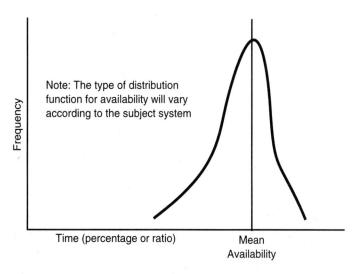

Fig. 1–2 Availability as a probability distribution function.

Reduced availability and production levels.   The plant does not have to be shut down to experience reduced availability. When many plant items fail, they do not shut down the plant. Nor do they always reduce its production level. However, the plant's characteristic availability has been reduced. This is because availability is a probability or expectation of performance that management hopes will continue over a period of time. Thus, its reduction is observable only "on paper." However, it is very real.

A simple example is a plant with two pieces of equipment. One is a spare. When one fails, the other is placed in service. Thus, the plant's real-time production level is not reduced. However, the probability of maintaining that level over a period of time is substantially less. This is because the expected availability performance has been reduced. This is so until the first item is returned to readiness or service. This phenomenon is demonstrated in Figure 1-3.

As a plant item fails, the probability distribution function for the physical availability to perform takes a less attractive shape and is positioned to the left. Other than the presence of a failed item, the reduction is not visible in plant-level performance. However, the plant's real-time probability distribution for availability no longer matches the expected optimal availability performance. Thus, the expectation of meeting plant production plans is now lower.

Alternately, a failure may result in a reduced production level in which availability at the desired production level is lost even though the plant continues to produce. A simple example is a plant with the same two pieces of

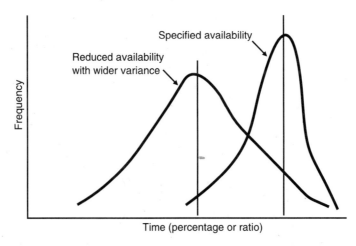

Fig. 1–3 Reduction of availability results from the failure of plant items.

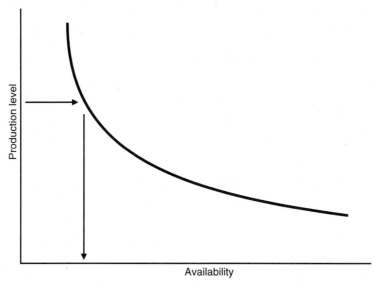

Fig. 1–4 Simple vision of the relationship of availability, reliability and maintainability.

equipment. In this case, however, assume both must work to achieve the desired production. When one fails, availability is lost when the plant is operating at full capacity.

This example reveals another point. To view the plant according to whether or not there is production does not accurately describe or measure success in a way that can be designed and managed. Instead, there are different probability distributions for each level of possible production. This is shown in Figure 1-4. As the plant is operated at lower levels, the expected availability is greater.

This point must not be missed. To say, for example, that a plant is operating at 97 percent of service capacity is not meaningful. A closer investigation may reveal that the result was achieved because plant performance is lower than acceptable. In other words, its availability performance may actually be poor although its service factor is high.

Availability and service factors compared. There is a tendency in business to equate the service factor and similar measures to availability. The discussion of availability as a probability distribution suggests that this is not possible for the following reasons:

- Service factor concepts deal only with what has happened. Availability is concerned with what is expected to happen. Thus, an important part of availability engineering and management is managing phenomena that cannot otherwise be seen and felt at the plant level until they result in reduced production.

- The service factor is often applied to measure aggregate performance. The facility is seen as a condition of "up or down." By comparison, availability is concerned with multiple, predefined acceptable levels of performance in an overall result. As a composite description of acceptable performance, these would be points along the curve of Figure 1-4.

Thus, the service factor is a simplistic measure of performance. It is probably fair to say that its simplicity reflects a history in which availability performance has not been treated with the sophistication it warrants.

Availability, Reliability, and Maintainability Defined

The heading of this section could easily be "Reliability with Maintainability Equals Availability." This is because reliability and maintainability are performance characteristics that combine to determine availability.

Three characteristics of availability.   Reliability, maintainability, and, therefore, availability are defined as follows:

- Reliability is the time between failures under planned operating conditions. It could be described as a period of continuous, trouble-free functioning.
- Maintainability is the time needed to maintain and return failed or shut down plant elements to service. By definition, it reflects planned working conditions, procedures, and resources.
- Availability is the fraction, ratio, or percentage of time that the plant or its subsystems are physically able to perform. Planned conditions are also inherent to its definition. However, they reflect the planned conditions for reliability and maintainability.

Reliability, like availability, involves equivalent concepts of probability. It is depicted as a probability distribution for time-to-failure or the need to shutdown the plant, a subsystem, or a piece of equipment. Defined operating conditions are the realm in which the probability distribution applies. If these change, so must the probability distribution.

Maintainability is also a probability distribution. It is the probability that the plant, subsystem, or piece of equipment can be restored to service in a period of time. Maintainability includes the expectation that the cost of maintenance tasks and their support requirements will fall within some range.

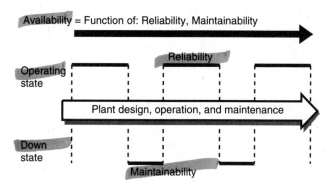

Fig. 1–5 Simple vision of the relationship of availability, reliability, and maintainability.

Figure 1-5 provides a simple vision for how reliability, maintainability, and availability are related. A plant, subsystem, or equipment item will be able to operate for some period of time between down states. This is its characteristic reliability. It will then remain down for some interval of time as a function of its characteristic maintainability. Availability is a function of both reliability and maintainability characteristics.

Basic formula for availability. The relationship represented in Figure 1-5 between reliability, maintainability, and availability is mathematical. Equation 1-1 applies the times for reliability and maintainability in the calculation of availability as a ratio, fraction, or percentage of time.

$$A = \frac{R}{R+M} \; \frac{MTBF}{MTBF+MTTR} = A \quad (1\text{-}1)$$

(handwritten annotations: "only failure" "What about cycle time Reduction & Planned D.T. Reductions?")

Where: A = Availability as a ratio, fraction, or percentage of time.
  R = Reliability as the time between failures.
  M = Maintainability as the time to return to service.

Reliability: The Interface Between Plant Subsystems

An earlier section introduced three interrelated money-making subsystems of plant performance. Cast with respect to productive capacity they were as follows:

- Producibility and composition of the plant's product line.
- Physical availability of the plant to perform.
- The production process and its operations.

Product and process operation determine the operating conditions experienced by each plant item. Reliability is the characteristic of performance that connects those conditions to availability. Therefore, the reliability component of Equation 1-1 reflects forces, reactive environment, time, and temperature. A change of the plant product, production process, or process operation will change the time between failures (reliability) that is expected for each affected plant item. As a result, the expectation for availability must change.

There are two possible responses to the situation. If availability is specified, the plant's characteristic reliability and maintainability must be realigned. If that is not possible, management's expectations for availability must be revised. Alternately, the plant's products, production process, and hard design may be revised. The ultimate response may be a combination of all four possibilities optimized as a system.

These possibilities suggest that to unilaterally or blindly modify the plant's product, production process, or design can reduce overall business results rather than improve them. For example, there are two possibilities when a production process is debottlenecked. The additional production revenues may be less than the additional cost to maintain the associated availability necessary to serve the plant's markets. Alternately, the additional process capacity, when offset by reduced availability, may not be realized in productive capacity.

The Three Subtypes of Availability

So far, availability has been defined as a single performance characteristic. However, there are three subtypes of availability. They are as follows:

Inherent availability (A_i). This measures the availability to be expected when reflecting unscheduled (corrective) maintenance only. Associated logistic and administrative time is excluded. The former is the time it takes to acquire and deliver resources to the repair task. The latter is the time needed to plan and administer the repair task. Thus, the maintainability element in Equation 1-1 is computed only with active repair time.

Achievable availability (A_a). This includes both scheduled and unscheduled maintenance. The former is defined as a situation in which a plant, subsystem, or equipment item is periodically shut down for overhaul. In Figure 1-6, Figure 1-5 is extended to show inherent and achievable availability.

Logistic and administrative time is still excluded. Therefore, inherent and achievable availability assume a perfect support environment.

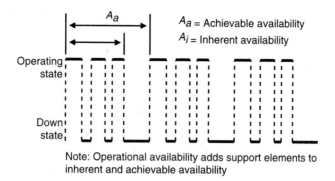

Note: Operational availability adds support elements to inherent and achievable availability

Fig. 1–6 Inherent and achievable availability.

The calculation of achievable availability still entails the principle of Equation 1-1, however, the calculation requires that the equation be extended to incorporate planned and unplanned shutdowns.

$$A_a = \frac{U}{U+S} \times A_i$$

$$= \frac{U}{U+S} \times \frac{R}{R+M}$$

(1-2)

Where: A = Availability as a percentage of time.
  R = Reliability as the time between failures.
  S = Time required for planned shutdown.
  U = Time between planned shutdowns.
  M = Maintainability as the time to return to service.

Operational availability (**A_o**). This includes logistic and administrative time in the computation of Equation 1-2. It reflects plant maintenance resource levels and organizational effectiveness.

Ultimately, operational availability is the bottom line. However, the following distinctions are necessary to design, measure, and manage integrated subgoals:

- Achievable availability fulfills the need to distinguish availability when planned shutdowns are included.
- Inherent availability fulfills the need to distinguish availability performance between planned shutdowns.
- Operational availability is required to isolate the effectiveness and effi-

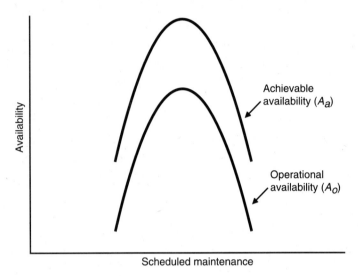

Fig. 1–7 Shape and location of achievable and operational availability

ciency of maintenance operations in availability performance. By comparison, inherent and achievable availability reflect production equipment configuration, selection, and design.

These definitions and distinctions lead to crucial recognitions. As indicated in Figure 1-7, they are as follows:

- The shape and location of the achievable availability curve in Figure 1-7 is determined by the plant's hard design.
- The point on the curve is determined by whether scheduled versus unscheduled maintenance strategies are selected for each failure. A goal of availability-based maintenance operations is to locate the peak and to operate at that level.
- Operational availability is the bottom line of performance. It is the performance that is experienced as the plant operates at a determined production level.
- The vertical location of the operational availability curve is controlled by decisions for resource levels and organizational effectiveness of maintenance operations. The previous definitions mean that it can never rise beyond achievable availability.

These factors have the following strategic implications:

- It is critical to know the location and shape of the achievable availability curve. Otherwise, it is not possible to determine what is reasonable and possible for operational availability and, therefore, plant production.
- If the upper curve is not known, manufacturing operations management may unknowingly attempt to achieve performance beyond what is possible. The result is like trying to fill a glass above its rim.
- Management must make strategic decisions for the long-term relative positions of the two curves. As plant production is increased over time, changing operating conditions will drive the upper curve down. Meanwhile, maintenance operation management must progressively move the lower curve upward to meet the demands of production. Eventually the two will converge. Additional availability can then be acquired only by a change in plant design.
- A conclusion may be drawn from these implications that availability engineering not done during the plant's original design project must ultimately be done for the existing plant. Otherwise, many of the current goals in industry to develop world-class maintenance operations are not possible.

Top-Level Factors of Reliability and Maintainability

Figure 1-5 shows that three organizational activities determine reliability and maintainability and their economics. They are plant design, operations, and maintenance functions. These activities can be jointly optimized if a plant is designed for availability performance in the capital project phase. Once the plant is built, operations and maintenance functions are the most flexible means to improve plant availability performance. However, some degree of business success is permanently lost when availability engineering is left out of the capital project.

Reliability, maintainability, and therefore, availability is the synthesis of many top-level factors and their application to each item in the plant as a system. Therefore, the permutations of the possible factors for achieving specified availability performance are immense.

Top-level factors.   Figure 1-5 suggests that top-level factors have two dimensions. One is reliability and maintainability. The other is decisions for design, operations, and maintenance. They are as follows:

Reliability

Is increased as the frequency of outages is reduced. Time between failures or shutdowns is increased.

Maintainability

Is increased as the duration of plant, subsystem or equipment down time is reduced.

Factors Driven by Design Decisions

Operating environment.

Equipment rated capacity.

Maintenance while the system, subsystem or item of equipment continues to function.

Installed spare components within an equipment item.

Redundant equipment and subsystems.

Simplicity of design and presence of weak points.

Accessibility to the work point.

Features and design that determine the ease of maintenance.

Plant ingress and egress.

Work environments.

Factors Driven by Maintenance Decisions

Preventive maintenance based on failure-trend data analysis.

Trend diagnoses and inspection of equipment conditions to anticipate maintenance needs.

Quality of the maintenance task (including inspection)

Skills applied to maintenance tasks.

How maintenance tasks are detailed, developed, and presented to the maintenance technician.

Quality of the system of maintenance procedures.

The probability of human, material, and facility resources being available to maintenance tasks.

Training programs.

Management, supervision, and organizational effectiveness.

Durability of handling, support, and test equipment.

Factors Driven by Operations Decisions

Reliability	Maintainability
Use of equipment relative to its rated capacity.	Organizational effectiveness as a factor in the troubleshooting process.
How spares are incorporated in normal process operation.	Organizational effectiveness and procedures to ready equipment for maintenance and startup.
Shutdown and startup procedures.	
Choices of raw materials.	

The first set of design decisions determines the difficulty of maintenance. The second and third sets are directly related to development and management of production and maintenance operations. All three sets involve decisions across reliability and maintainability.

Thus, the plant capital project provides an opportunity to design the top-level factors as a system. Once the plant is built, management must look to production and maintenance operations management to advance plant availability performance. The reliability and maintainability factors they control can be aligned to complement those that have become a "given." However, the previous discussion of operational availability showed that there is an upper limit to this strategy.

Goals of Availability Engineering and Management

The central goals for availability engineering and management are as follows:

- To collaborate with production process operations in the achievement of the maximum profitable productive capacity.
- To achieve the most profit-effective availability performance as part of plant productive capacity.

These are related. However, they recognize two purposes of availability engineering and management: maximally profitable productive capacity and associated availability performance. They are discussed respectively in the following sections.

First Central Goal: Productive Capacity

It was mentioned that a plant's productive capacity has three sources. Figure 1-8 shows that they are offsetting costs. Product and process opera-

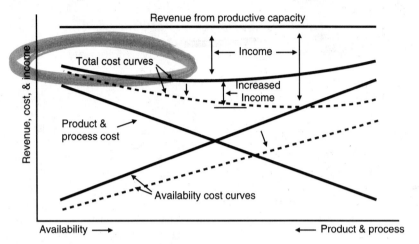

Fig. 1–8 Product and production process offset availability in cost of plant production capacity.

tions are offset against availability for a given level of productive capacity. The goal is to function at the bottom of the cost valley. This takes into account availability performance as a major determinate of maximally profitable productive capacity.

The importance of this relationship can be seen in the context of a break-even graph. As the balance is optimized, the cost curve of Figure 1-9 is shifted downward and rotated clockwise with respect to the revenue curve. Thus, business results are maximized. The figure shows the causes and consequences of rotated and shifted cost curves.

Second Central Goal: Profit-Effective Availability

The second central goal of availability engineering and management is to achieve the maximum profit-effective availability performance. There are three associated fundamental subgoals:

- To achieve management's specified mean and distribution for availability by methodical design.
- To optimize the three types of availability and associated reliability and maintainability factors as a system for achieving the specified distribution of performance.
- To achieve the most profit-effective life-cycle cost as a criterion of optimization.

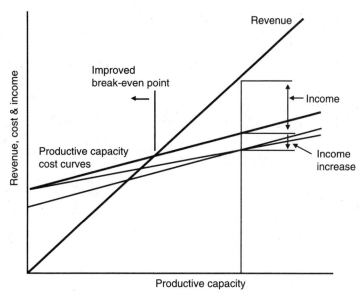

Fig. 1–9 Affect of optimal productive capacity on the financial break-even point and income

The third subgoal is demonstrated in Figure 1-10. Reliability and maintainability are economically interchangeable in the creation of income. This is true for each type of availability. As the choice is skewed toward one characteristic, the life-cycle cost of the specified availability is eventually increasedFailure to capitalize on the interchangeability of the previously introduced reliability and maintainability factors may lead to excessive life-cycle cost. This means that the individual factors behind both are maximized rather than correlated. Plant availability and, therefore, the resulting normal cost of overall productive capacity will grossly exceed what is required and plant income will be reduced. Without availability engineering and management, this outcome may easily be the case..

The first two subgoals suggest the necessity to connect reliability and maintainability by design rather than default. However, there are many ways to achieve this. Figure 1-10 shows that the third subgoal is concerned with finding the most financially attractive design for availability performance.

The importance of an optimal economic balance is related to the first central goal. The cost curve for availability in Figure 1-9 is shifted downward and rotated clockwise when reliability and maintainability are optimized as shown in Figure 1-10. Furthermore, the break-even point of Figure 1-9 is shifted to the left. This affects the cost curve of Figure 1-8 for plant productive capacity. Income increases as the total life-cycle cost curve develops a deeper valley.

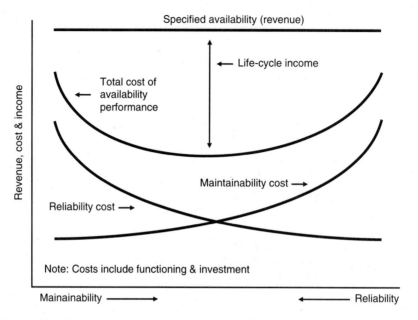

Fig. 1–10 Profit-effective choices for reliability versus maintainability to achieve a specified availability.

Goals When the Plant Design is Fixed

An availability design team may be faced with a plant in which the production process, configuration, and equipment has been substantially or completely fixed. This may be the case when the design of the plant has progressed to the detailed design phase or when the plant already exists.

This does not negate the necessity and possibility of availability engineering and management. Some of the top-level factors for reliability and maintainability have become "givens." This is more often the case for reliability than for maintainability.

However, availability engineering and management still has two objectives.

• To determine the expected reliability from the given plant design. Expected availability is the issue for an existing plant.
• To design the flexible top-level factors to complement the "given" elements of reliability and maintainability. Existing plants will require alignment rather than design. The greatest number of still-flexible factors will be associated with maintenance operations.

Business and Production System Optimization and Plant Suboptimization

The chosen balance may not rest in the valley of the plant's life-cycle cost curve. There may be external social and economic goals that prevent it. For example, the owner may have several plants that produce the same product, or the plant may be just one facility in a system of upstream and downstream production facilities.

Optimal design and management is still the goal. However, it should reflects a global rather than local optimum. The availability discipline addresses these external goals.

Summary

The attributes of availability and the goals that influence its design and management have been defined in this chapter. The next chapter explores the immense financial ramifications of availability performance on corporate, operating company, and plant-level business results.

Bibliography

Blanchard, Benjamin S. *Logistics Engineering and Management*. 4th ed. Englewood Cliffs, N. J., Prentice Hall, 1991.

Bloch, Hienz P. and Geitner, Fred K. *An Introduction to Machinery Reliability Assessment*. New York, Van Nostrand Reinhold, 1990.

Cleveland, E.B., Jarrett, A.A., and Morrison, C.B. Assessment of Methods for Implementing Availability Engineering in Electrical Power Plants. Electric Power Research Institute. EPRI NP-493. May 1977.

Lamb, Richard G. Approach to Making Maintenance Operations the Strategic Resource for Availability Performance in Asset Utilization. Technical paper to 1993 NPRA Maintenance Conference.

Williams, Gerald P., Hoehing, William W., and Byington, Robert P. Causes of Ammonia Plant Shutdowns: Survey V. Presented to AIChE Ammonia Safety Symposium. Boston, Mass. August 1986.

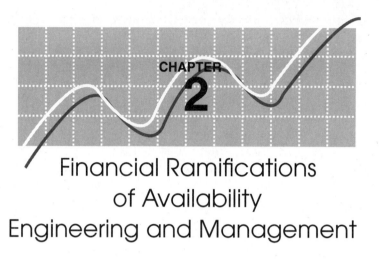

Financial Ramifications
of Availability
Engineering and Management

Introduction

A central goal of availability engineering and management is to design the plant for optimal life-cycle income and productivity of capital and working assets. This has important implications for the owner's short- and long-term financial statements; return on investment; earnings per share; share value; and the need, capacity, and cost of raising debt and equity capital.

The purpose of this chapter is to explore the financial ramifications of availability performance. It does this from various financial perspectives as measures of business success.

The following financial analysis and management concepts will be used:

- Life-cycle cash-flow profiles.
- Payout and discounted cash-flow analysis.
- Life-cycle income analysis.
- A return on investment (ROI) model for the integrated analysis and design of income and productivity of assets.
- Financial break-even point and operating leverage are used to demonstrate that a small percentage of improvement in availability and its associated cost has a substantially greater percentage of improvement in plant income.

Each of these items is discussed in the following sections. However, the purpose of this chapter goes beyond a discussion of the financial ramifications of these concepts.

The basic design phase of availability engineering (See Part 3) includes the development of a financial model. Because a central purpose of the discipline is to go beyond cost-benefit and discounted cash-flow analysis, these financial performance concepts are incorporated in the model. This chapter explains why these financial concepts are important to availability design and management.

Consequences for Life-Cycle Cash Profiles

It may be helpful to first describe the possible consequences of availability performance on plant cash-flow profiles. These include:

- Revenue profile.
- Manufacturing (operations and maintenance) cost profile.
- Capital expenditures profile.

These profiles are shown in Figure 2-1 and how they are influenced by availability engineering and management is discussed in the following sections.

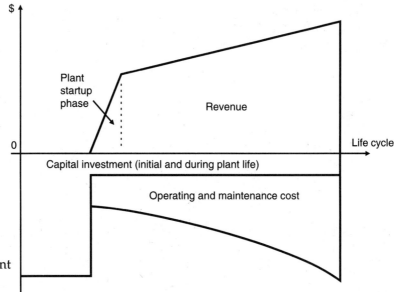

Fig. 2–1 Simple plant life cycle cash flow profiles.

Revenue Profile

The life-cycle revenue profile is affected in several ways. They are as follows:

- Achieving management's specified availability performance.
- Being a low-cost producer in the plant's markets.
- Reducing the time necessary to achieve full commercial production.
- Continually improving performance through living design.

Achieving specified availability. The design of reliability and maintainability to meet availability performance targets will shift the revenue profile upward. This shift is the result of two factors:

- Production is physically possible a greater percentage of the time.
- Average production is closer to the rated capacity a greater percentage of the time.

Becoming a low-cost producer. Being the low-cost producer is especially important for commodity products. The ability of the plant to be such a producer shifts its revenue profile further upward. The shift, which reflects optimal life-cycle investment and maintenance cost profiles is the result of achieving the cost valley for availability performance shown in Figure 1-10.

Why this is so is shown in Figure 2-2. The following points are made:

- Each step of the curve represents a plant and its respective cost structure in terms of product unit-cost.
- Each business cycle has an associated demand. As business conditions worsen, the vertical line representing the level of demand moves to the left.
- The umbrella product unit-price in each business cycle is generally set by market forces to reflect the high-cost producer. This producer is the one whose unit cost as a function of its overall cost structure intersects the demand line.
- Plants to the right of the demand line are shut down until demand increases and prices improve in response to increased demand. A study of ammonia plants had a euphemism for such a case: "Shut down for inventory control."

Thus, the plant owner's objective is a cost structure positioned as far as possible down the plant's curve below the dynamic umbrella price. The life-cycle revenue profile in Figure 2-1 is shifted upward by this accomplishment.

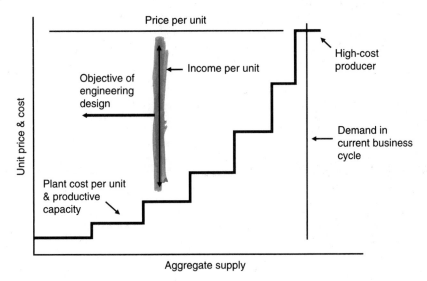

Note: As competitors move down the curve, the width of their steps
(productive capacity) will increase. Movement upward by one
competitor ia accelerated by the downward movement of another

Rule: For a level of demand, price of a commodity
generally equals the cost of the high-cost producer

Fig. 2–2 Objectives and benefits of being a low cost producer.

The location on the curve will not just affect income per unit of production. In some business cycles, it will be the difference between generating revenues or being shut down for "inventory control."

The curve also suggests that there are considerable strategic consequences of failing to advance a plant's availability performance. Plants whose performance is moved down the stepped curve push upward those whose performance is allowed to stand still. Worse, this upward movement is accelerated. The width of the steps represents productive capacity. As the width for advancing plants increases, the position of nonadvancing plants is pushed even faster up the curve.

Revenue performance of full commercial production. Well designed maintenance operations will improve early revenue performance. The leading edge of the revenue profile shown in Figure 2-1 is shifted forward and is rotated counterclockwise. This is the result of maintenance operations designed and appropriately deployed to deal with the special challenges of startup and early commercial production.

Continual improvement through living design. Some products of the availability engineering process are its living design functions and systems. The availability scheme is not designed, deployed, and then placed in archives. Instead, its work products are assigned to various functions and installed in computer systems for continual use. Uses include availability analysis and design initiatives as well as normal corporate, operating company, and plant functions.

The result is that activities associated with living design bend the revenue profile counterclockwise over time. For some facilities, this may mean slowing the rate of revenue decline in the plant's later life.

Operation and Maintenance Cost Profile

The life-cycle operation and maintenance (manufacturing) cost profile is affected by availability engineering and management. The discipline affects these functional cost segments as follows:

- It has an indirect effect on the productivity of the production process.
- It determines appropriate resource levels, procedures, and functions for profit-effective maintenance operations.

Production operations.   Design and operation of the production process determines both the input to and waste from producing the plant's products. Material, energy, labor, administration, and management costs make up the total-life cycle cost profile.

Total production cost reflects how closely the process can operate according to its design. The process control system is fundamental to achieving this goal. Accordingly, availability engineering and management indirectly affects the production component of the plant functioning cost. This is because the control system is designed and managed to be appropriately available to perform.

The value of control system availability to the revenue profile is also significant. A study of ammonia plants found that control system failures were a significant cause of plant shutdowns [Williams, Hoehing, and Byington, 1986].

Maintenance operations.   The other major component of the total plant functioning cost profile is maintenance operations. This is the case for any equipment-intensive production process. It is especially the case when production personnel are not elements in the process, but only enable and respond to its needs. Next to raw materials, maintenance activities will be the greatest cost of production.

An activity-based cost analysis at Union Carbide Corporation's Texas City complex revealed the significance of maintenance costs [*Chemical Week*, 1993]. The study found that maintenance accounted for 30 percent of the plant's total work hours. Production operations were the next greatest application of work hours, however, they accounted for only 12 percent of the total number of hours. It was not apparent if the classification of maintenance hours fully included the maintenance related activities of operating, operating support, and general administrative personnel. If not, the number of hours is conservative.

The availability discipline minimizes maintenance costs because it avoids the following possibilities.

- Inadequate management and field resources leading to purely tactical, reactive maintenance actions. These actions may be directly or indirectly expensive. They may be ineffective or even destructive. The result may be excessive maintenance costs and other significant hidden losses.
- Excessive resources leading to a high ratio of costs with respect to a level of income.
- Inappropriate maintenance strategies and poorly developed maintenance procedures and functions also leading to both of the previous results.

This suggests that an optimal availability scheme must precede the determination of resources. It also implies that the use of people, material, and maintenance equipment must be managed effectively in that scheme.

Continual improvement through living design. We have already discussed the effect of the living design on the revenue profile. Likewise, the operating and maintenance cost profile in Figure 2-1 will be bent counterclockwise over time. For some plants, this may reduce the rate that maintenance operation costs increase as the plant ages.

Capital Investment Profile

Capital investment includes the costs of initial plant design and construction. It also includes sporadic capital projects during the plant's productive life.

Justification of investment decisions. The justification of capital expenditures is a fundamental issue of plant design. Thus, whether availability engineering increases or decreases capital investment is not as important as whether the investment is appropriate for the plant's optimal availability performance in productive capacity and business operations.

The availability discipline offers the following possibilities:

- The justification of functional redundancy in the plant's configuration can be confirmed. Availability analysis plays the same role in justifying equipment with high or low reliability and maintainability.
- Availability engineering may lead to the discovery of enhancement opportunities during the design phase rather than later in the plant's life. This is a primary goal of the discipline because enhancements made during the construction phase are usually less expensive than those discovered and installed later in the plant's life cycle. Furthermore, enhancements made later in the plant's life will have been preceded by permanently foregone income opportunities.

Payout and Discounted Cash-Flow Analysis

Life-cycle cash flow was the subject of Figure 2-1. Time-to- payout and discounted cash-flow analyses measure the net result of the profiles. They treat each plant design decision as a choice between alternate, but equivalent, investments. Availability engineering and management has dramatic implications for the results of both calculations.

Definition, Purpose, and Comparison

Payout analysis. Payout analysis calculates the time required for income to recover its associated investment. It does not reflect the time value of money. Nor does it show the consequences of the investment after payout. Therefore, its value is limited to the assessment of the investment up to the time of its payout.

Decisions based strictly on payout analysis are not good management. However, the method is still pertinent because it is useful in establishing parameters that reflect the constraints and goals of capital and liquidity. Examples are conditions like the following:

- Constraints on external capital.
- A policy to invest within the owner's liquidity.
- Policies to maintain certain dividend levels, equity-to-debt ratios, weighted cost of capital, *etc.*
- Constraining cash position and debt commitments.

Discounted cash-flow analysis. Financial theory holds that the continuing purpose of management is to increase the organization's net present value. Discounted cash-flow analysis determines which investment options best fulfill that responsibility. The analysis calculates net present value and internal rate of return of plant investment, revenues, and operations and maintenance (manufacturing) cost profiles.

All plant- and element-level design decisions will variously affect the cash-flow profiles. Thus, discounted cash flow is an important tool for evaluating choices. Each choice can be weighed by:

- Comparison of solutions.
- Testing a solution against a hurdle rate of return.

Unlike payout analysis, discounted cash-flow analysis is based on the time value of money over the life of the investment. However, the analysis can be misleading because it gives greater mathematical weight to the short-term. It does not properly reflect the importance of increased availability and decreased operations and maintenance costs. This may result in equipment costs being given grossly disproportional importance in plant design.

Sensitivity to Availability Performance

Time-to-payout and discounted cash flow are sensitive to the performance made possible by availability engineering and management. This is because:

- Plant design meets, but does not exceed, specified availability. This means that management is not unconsciously investing in productive capacity it did not strategically intend to create. As a result, time-to-payout is decreased and net present value is increased almost as much as the avoided unjustified capital expenditure. The rate of return is also substantially increased.
- Income is increased. This is the result of developing the most profit-effective maintenance operations as part of the plant's availability scheme.

The payout calculation, however, essentially highlights a minuscule period of time relative to the total life span of the plant. The mathematics of discounted cash-flow analysis also gives great weight to investment and shorter-term incomes. Therefore, the result of optimal maintenance operations does not affect either method to an extent that they are significant to the plant's lifetime financial performance.

- Time to reach full commercial production may be shortened. This is a goal of availability engineering and management during the start-up phase. Payout and discounted cash ow are sensitive to this condition.

Availability Design for Maximum Payout and Discounted Cash-Flow Effect

It an earlier discussion it was suggested that payout may be associated with capital management constraints and goals. Discounted cash-flow analysis also supports these goals.

If these goals are extremely important to an organization, management should establish investment minimization as a high-ranking design criterion. The reduced capital investment will be offset by the increased cost of lifetime maintenance operations. The plant and its equipment should be designed for less redundancy and durability and a lower level of maintainability. In turn, the maintenance operation must sustain higher human and material resource levels. Management requirements and costs will also increase.

The objective of this design is still to increase the time between failures and to reduce the time to return to service. However, the specified availability performance is achieved with a greater stress on working, rather than capital, assets.

Availability design must determine whether there are parameters for these capital management goals and identify the logic behind them. This is critical because a design optimized on payout, net present value, and rate of return may be inconsistent with more holistic performance criteria.

Purpose of Payout and Discounted Cash-Flow Analysis in the Financial Model of Availability Design

As mentioned, the purpose of this chapter is to explore the financial ramifications of availability engineering and management. It is also to identify how the availability scheme must be designed with respect to different financial parameters of overall business performance.

Payout and discounted cash-flow analysis can help an organization to achieve these goals because they provide the ability to screen and compare alternative designs. However, their sensitivity to integrated financial ramifications is limited. Worse, they are potentially misleading. Thus, these methods must be used carefully in financial modeling (basic design phase) for availability. They should be tempered with the analysis concepts described in the following sections.

Life-Cycle Income Statement Analysis

Cash flow profiles, payout, and discounted cash-flow analysis are fundamentally weak in one major respect. They do not easily illustrate and measure financial parameters against which organizations are normally judged. More specifically, they do not show how the investment, functioning costs, and revenue they create appear as income. Thus, the ramifications to owner income statements over the plant's life time are not very apparent.

Simplistic View of Life-Cycle Income Measurement

A simple life-cycle income statement is presented in Figure 2-3. Although simplistic, it is still an accurate representation of the financial issues discussed in this section.

The statement is simple for several reasons. The shape, fluctuation, and relative magnitude of revenue and operating and maintenance expense profiles vary according to industry. This is also true for the shape of the profile of investment translated to depreciation expense. These differences are due to the various accounting methods used and to the nature of each industry.

The figure shows the universal attributes of income that are important to availability design and management. They are as follows:

- Capital expenditures and engineering costs do not appear until commercial production begins. They appear over a lengthy period as a depreciation expense. This is very different from the cash-flow vision of financial consequences.[1] Thus, a perception that the owner is automatically served by the least amount of investment rather than profit-effective life-cycle availability performance is belied.
- Capital expenditures and functioning costs are matched to the revenues they ultimately produce. This is the meaning of income as compared to cash flow. Without this matching, business performance would be difficult to design and measure.

---

[1] Capital expenditures flow into an asset account as they are incurred.

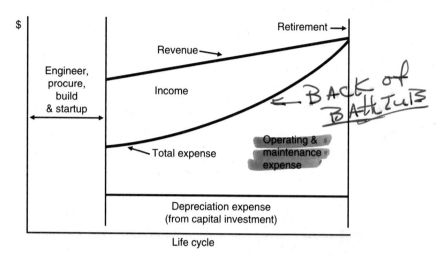

Fig. 2–3 Simple life cycle income statement profiles.

- The expense and revenue profiles will ultimately cross and the plant will cease to be financially viable.
- Capital investment is much less significant in the calculation of income than total direct and indirect operations and maintenance expenses.

These conditions suggest that a goal of availability design and management is to maximize the income slice of Figure 2-3. This goal is expressed in this book as it pertains to *profit-effective* rather than *cost-effective* results. In fact, in the business world, the first term was originally intended to mean the latter. However, there is a tendency in business discussions to be concerned with saving, rather than making, money.

Affect of Availability on the Life-Cycle Income Statement

Availability design and management is profoundly concerned with the attributes of the income statement. The results of availability performance are as follows:

- The revenue profile is shifted upward. However, this shift must actually increase income. This is achieved by designing the plant to attain the availability associated with that profile.

[handwritten annotations:]

BACK of BATHTUB

The Bath-Tub Curve is flattened & Extended By Cost effective Maint. Policies & Procedure

Only if the market to sell the product is available.

- The revenue profile is not matched against unjustified functioning and depreciation expenses. This is the case when designing without method results in more availability than the market can bear. As a result, the plant's ability to perform may not be consistent with its markets and strategic purpose.
- The revenue profile may be shifted further upward. This is the indirect result of being a low-cost producer. Availability performance is fundamental to this position. This point is demonstrated in Figure 2-2. Higher-cost producers can be shut down during some business cycles while the remaining lower-cost producers continue to have a market.
- Availability engineering and management minimizes the combined operations, maintenance, and depreciation expense profile. Figure 1-9 illustrated the importance of striking a balance between reliability and maintainability when designing the plant. This balance is reflected by a valley in the total life-cycle cost.
- A consequence of these design goals is to widen the income slice.
- These goals will also extend the life span of income. Delaying the intersection of revenues and total expenses is an important goal of availability engineering and management. Attention to this goal begins with initial plant design.

  True; But maint can also contribute

- The revenue profile will be bent counterclockwise over time. Meanwhile, the total expense profile is bent clockwise. This further widens the income slice and extends economic life. This condition is the result of the living design that was established as part of availability engineering and management. Thus, it serves as a platform for the continual testing, evaluation, and improvement of the plant's most profit-effective availability performance.

Purpose of Life-Cycle Income Analysis in the Financial Model for Availability

Life-cycle income analysis must be included in the financial model for availability design and management because there is a financial limit to shifting life-cycle cost from investment to maintenance operations. Overall financial performance will cease to improve and then begin to fall off. Payout and discounted cash-flow analysis methods will not easily recognize the point at which this occurs. Life-cycle income analysis does not suffer from this weakness.

As mentioned, the financial model is structured in the form of financial statements. The income statement is one of them. The previous discussion showed its importance to the model.

Return on Investment Analysis

Return on investment is computed as accounting period income divided by total assets. However, the computation of return on investment for the current period is not the primary point of interest of availability engineering and management. It is more concerned with the financial optimization of availability elements such that return on investment is maximized over the plant's lifetime.

Return on Investment Model

Purpose of the model. A financial model for return on investment links, views, and evaluates the integrated financial elements of availability performance. These relationships are both within and between the income statement and the balance sheet.

The ROI model demonstrates complex, integrated financial ramifications of availability engineering and management. Thus it is a powerful design tool for measuring the availability scheme with respect to return on investment.

Structure of the financial model. The financial model is the subject of Figure 2-4. It is a combination of financial accounts and measures mathematically related in business results through return on investment.

The respective sides of the model are related to the income statement line items and balance sheet accounts as follows:

- The income statement side is concerned with the effect of availability performance on income and income margin.
- The balance sheet is concerned with the productivity of assets and how well the availability sche;me utilizes the owner's fixed capital and working assets.

Working assets are cash, near cash, parts and materials inventories, and accounts receivable. Accounts payable is included in working asset management. The availability scheme may substantially affect all but accounts receivable.

Performance efficiency!

An added dimension. The ROI model adds another dimension to availability design: the productivity of assets. This is an important addition because without it, the plant's financial performance may not reflect the productivity of working and capital assets. Therefore, maximizing return on investment is a useful means for optimizing income and productivity of assets as a system of performance.

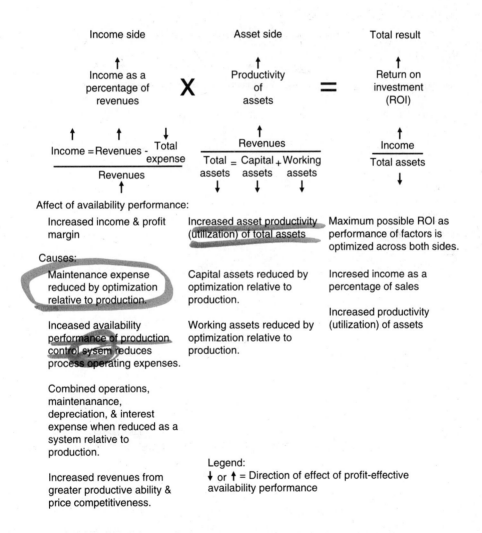

Fig. 2–4 Affect of availability performance on the elements of return on investment.

Affects of Availability on Income Statement Side Elements

Availability engineering and management affect the income side of return on investment as follows:

TPM

- Operating expense is part of total expenses. It is reduced through the availability performance of the process control system. As a result, the production process can better sustain operations at its intended most economic input/output ratios.
- Maintenance operations expense is also a large part of total expenses. It is reduced by finding the cost valley for a specified availability. This includes direct and indirect costs.
- The combination of the operations, maintenance, depreciation and interest expenses is reduced when optimized as a system. This was described in Chapter 1 as one of the two central goals of availability engineering and management. The result is represented by the total cost valley for productive capacity shown in Figure 1-7.
- Increased revenues result from achieving management's desired availability. They may be further increased by the price competitiveness created by the previous factors.
- The increased income and its margin results in greater income as a percentage of revenues.

Affects of Availability on Balance-Sheet Side Elements

Availability engineering and management affects the balance sheet side of the return on investment as follows:

ROFA?

TRUE; But APPEARS TO BE TO CLOSED ON EXISTING PLANTS

- The capital assets (investment) balance may be reduced. This occurs if unjustified redundancy and costly equipment is discovered by the availability engineering process.
- The capital asset balance may also be increased by the availability engineering process. This will be the case if the evolving plant design is found to be inadequate or suboptimal with respect to specified availability performance. However, such a discovery is countered many times over during the plant's lifetime by increased revenues, decreased total expenses, and decreased working assets.
- Working asset requirements will be made consistent with an optimal availability scheme. This is important because the capacity to calculate a provisioning level is significant to the overall working assets. Important

working assets are spare/repair parts and materials. The cash balances that support maintenance operations, and human, parts, material, and other support elements are also important working assets.

These are good examples of the integrated effects availability design and management must consider. Reduced provisioning concurrently reduces management expenses. Thus, the income side of return on investment will be further enhanced along with the productivity of assets.

- The result of these factors is a higher productivity of assets. This means that there are optimal combined capital and working assets levels associated with a forecasted level of revenues. *ISN'T this REALLY the GOAL of TPM?*

Return on Investment: The Synthesis of Both Sides

Ultimately, both sides of the model are synthesized as a single measure. Income as a percentage of sales (a percentage) and productivity of assets (a ratio) are multiplied to compute return on investment. As both sides of the model are improved by profit-effective availability performance, so is return on investment.

However, there is a greater goal. It is possible that decisions for one or both sides may not be optimal. The goal is an overall scheme that maximizes return on investment.

Purpose of Return on Investment Analysis in the Financial Model for Availability

The value of availability engineering and management is stressed by each of the model elements. Thus, it is fundamentally important that plant availability include, along with income, a design for the productivity of assets and the maximum return on investment.

The return on investment model is an important part of the financial model of availability performance because it provides a return on investment readout to assist the design team in the decision making process. It provides an important advantage because it views profitability beyond life-cycle income. Thus, it an important tool for determining the mixture of capital investment and functioning expense that would produce the optimal business performance over the plant's life cycle.

The Break-Even Point and Operating Leverage Analysis

The financial concepts of the break-even point and operating leverage demonstrate an important phenomenon. Improvements in availability related revenues, investment, and maintenance operation expenses have a disproportionately greater consequence for income. This is illustrated by the break-even point diagrams in Figures 2-5 and 2-6.

The Break-Even Point and Income

Figure 2-5 and Figure 2-6 show that a plant has a cost structure. Its components are direct and indirect, fixed and variable. The structure determines when during the accounting period revenue will exceed expenses. Beyond that point, the plant produces income rather than loss.

There are three consequences of profit-effective availability performance. They are as follows:

- The cost curve in Figure 2-5 shifts downward and rotates clockwise. The figure depicts these movements separately, but they are actually a

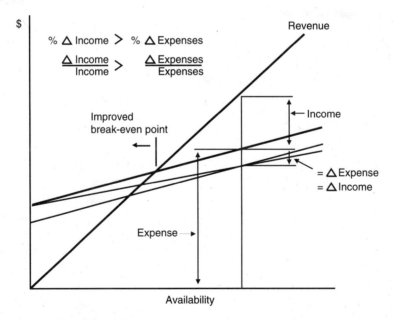

Fig. 2–5 The leveraged consequences of life cycle costs on income.

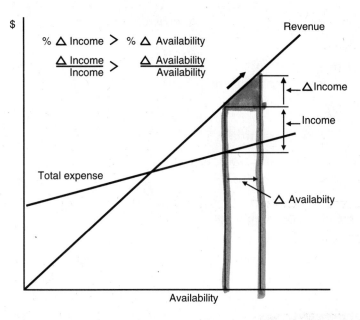

Fig. 2–6 The leveraged consequences of increased availability on income.

combined result. The final resting place of this combined result reflects the optimal cost structure for availability performance.

- The break-even capacity utilization point, in turn, shifts to the left.
- Figure 2-6 shows that there is also an upward movement along the revenue curve. This is the result of achieving a specified availability performance and greater cost competitiveness.

Thus, income first occurs at a lower production level. Income for subsequent periods is concurrently greater.

Operating Leverage

The full significance of the previous scenario is demonstrated by operating leverage. Operating leverage is defined as a percentage of change in sales and cost structure that results in a greater percentage of income. It is expressed as a ratio.

This increase is expressed geometrically in Figures 2-5 and 2-6. Note that:

- A shift or rotation of the expense curve generates some percentage of reduction in expenses for the period. A substantially greater percentage change occurs in income.
- A percentage movement further along the curve has the same result. There will be a substantially greater percentage change in plant income.

Both consequences of operating leverage are generated by availability engineering and management.

The Relationship of the Break-Even Point to Operating Leverage

The location of the break-even point is mathematically related to operating leverage. Some industries have such competitive markets and high cost structures that normal production is typically just above the break-even point. These industries have very high operating leverage. Thus, achieving basic competence in availability engineering and management will have immense consequences.

This relationship is a two-edged sword, however. Design and management decisions without a tie-back to availability performance can just as easily and unknowingly reduce overall financial performance. To ignore methods for developing profit-effective availability performance is not just a decision to "do nothing." It is a default decision to accept reduced business results. This will be the case even if maintenance operations are efficient, effective, and are continually improved.

The Purpose of Analysis in the Financial Model for Availability

The break-even and operating leverage concepts are built into the often mentioned financial model for availability design. This is necessary since the break-even point is a function of fixed and variable components in the availability cost structure. The capacity to determine the optimal position between these extremes is not a feature of the previously described financial performance analyses and measurement methods. Thus, it is important to quantify the break-even point and the degree of operating leverage that is inherent to the plant.

Summary

The financial ramifications of availability engineering and management were discussed in this chapter. It was shown that accounting rules, the relative magnitude and shape of the financial profiles, and the significance of performance measures vary according to industry, type of plant, stage of the plant's life cycle, stage of the product and market cycle, etc. However, these relationships and ramifications are completely relevant to any operation. Thus, they also define the fundamental design criteria and parameters for availability design and management.

Bibliography

Morris, Gregory. "An Old Plant Pioneers New Maintenance Methods." *Chemical Week.* June 8, 1993.

Thuesen, Gerald G. and Fabrycky, W.J. *Engineering Economy.* 8th ed. Englewood Cliffs, N. J., Prentice Hall, 1993.

Weston, J. Fred and Brigham, Eugene F. *Managerial Finance.* 10th ed. Hinesdale, Ill., The Dryden Press, 1993.

Williams, Gerald P., Hoehing, William W., and Byington, Robert P. Causes of Ammonia Plant Shutdowns: Survey V. Presented to AIChE Ammonia Safety Symposium. Boston, Mass. August 1986.

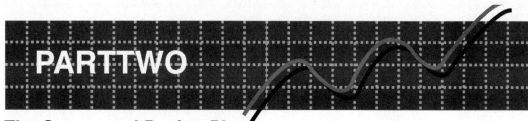

PARTTWO

The Conceptual Design Phase

CHAPTER 3

Availability Engineering and Management and the Conceptual Design Phase

Introduction

Traditionally, the conceptual design phase includes, in some form or approach, the following:

- Needs analysis, feasibility studies, and the ultimate investment decision.
- The evaluation and selection of a production process.
- A detailed engineering description of an associated plant scheme.

The goal is a production process and plant scheme that can achieve specified production parameters such as rate, heat, pressure, *etc.* However, this is only part of the full plant design problem. This traditional scope does not take into account the need to design the plant for its physical availability to perform the production process. *uptime?*

The traditional design methods have historically been subjected to a disciplined engineering methodology. But although the many aspects of availability engineering and management have sometimes been recognized as crucial design methods, they have not been subjected to an equally disciplined process.

This is remedied by expanding the conceptual design phase. Phase deliverables are added for availability design. They form the vision, means, and measure for how the plant will be made physically available to perform in the most profit effective way. Those deliverables are the subject of this chapter.

Requirements for Availability Design in the Conceptual Phase

The availability discipline introduces the following elements to the conceptual design phase:

- Policies for availability engineering and management are established by management. They are applicable to the plant project phases and its remaining productive life.
- An availability concept is developed. It is a top-level description and set of policies, criteria, and parameters for availability performance.
- A maintenance operation concept is developed along with the availability concept. It too is a top-level description and set of policies, criteria, and performance parameters. However, its domain is to establish how the plant will be maintained throughout its producing life.

Approach to This Chapter

Traditional and availability phase activities and their work products should be approached as inseparable processes. In the past, the phase activities for availability have often been treated as independent design processes, if treated at all. This chapter addresses both of these approaches. They are presented in the order in which they would occur in a comprehensive conceptual design. *Comprehensive* is defined as what is necessary to fully design for *how the product is produced*.

Figure 3-1 shows the project deliverables of the conceptual design phase. The deliverables are coded as "TC" and "AC." This denotes *traditional* and *availability* deliverables during the *conceptual* design phase.

Each deliverable is described in the following sections. The availability deliverables are described in depth and supported with detailed flow charts. The traditional deliverables are touched upon briefly. The descriptions includes the phase process, but focuses chiefly on the availability engineering deliverables.

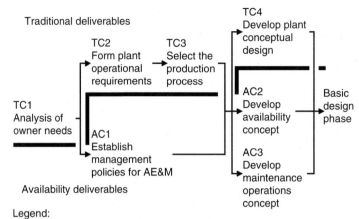

Fig. 3–1 Availability engineering deliverables of the conceptual design phase

---

TC1. *Analyze Owner Needs*

The conceptual phase begins with identification and analysis of the plant owner's basic needs. The most likely recognized need is to achieve or protect some strategic position. Management expects that the organization's business results will be enhanced. This initial need exists for new plant development projects as well as for initiatives to improve existing ones.

The identified need is refined by marketing and strategic studies. They add definition to the desired strategic position, provide an estimate of long-term business consequences, and determine necessary resources and commitments.

Therefore, the purpose of owner needs analysis is to:

- Establish that there is an actual need.
- Develop a definition of the need.
- Formulate appropriate strategies.
- Make final choices and decisions for plant investment.

The ultimate work products of the owner needs analysis are important to availability engineering and management. They help define the strategic nature and value of availability performance in the subject plant.

AC1. Establish Management Policies for Availability Engineering and Management

Definition and Purpose

Management must directly or indirectly establish policies for approaching the capital project including those associated with availability performance. The purpose of this deliverable, then is to identify general policies and translate them to policies applicable to the availability discipline. The resulting policies are not limited to the design for availability in the capital project; they also mandate availability management throughout the plant's producing life.

Establishing Policy

The requirements for formulating policy for availability engineering and management in the capital project and other initiatives (Figure 3-2) are the subjects of the following deliverables.

AC1.1. Identify pertinent regulations and industry initiatives. Regulations and industry initiatives that lead to policy decisions for availability must be searched out and evaluated. They often rely upon the availability engineering and management discipline to achieve their recommendations and mandates.

For example in some industries, safety regulations and industry initiatives stipulate requirements for the mechanical integrity of process and safety equipment. This means that the availability discipline must include a design for managing safety-relevant plant items.

AC1.2. Identify pertinent owner policies and initiatives. In most organizations there is a plethora of pertinent corporate, operating company, and plant-level policies and initiatives. Some of these may have immense ramifications for business success. Thus, this task is to establish the need for definitive availability policies.

The following are examples of such policies and initiatives:

- A company has mandated that all future projects include designs for constructability, operability and maintainability. Chapter 1 showed that the last requirement is actually a mandate for availability engineering and management.

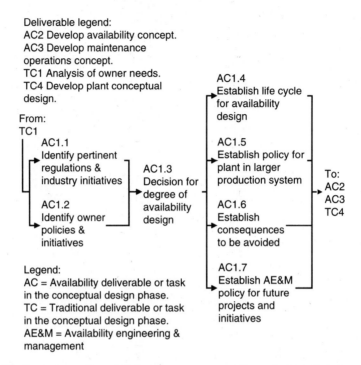

Deliverable legend:
AC2 Develop availability concept.
AC3 Develop maintenance operations concept.
TC1 Analysis of owner needs.
TC4 Develop plant conceptual design.

From:
TC1

AC1.1 Identify pertinent regulations & industry initiatives

AC1.2 Identify owner policies & initiatives

AC1.3 Decision for degree of availability design

AC1.4 Establish life cycle for availability design

AC1.5 Establish policy for plant in larger production system

AC1.6 Establish consequences to be avoided

AC1.7 Establish AE&M policy for future projects and initiatives

To:
AC2
AC3
TC4

Legend:
AC = Availability deliverable or task in the conceptual design phase.
TC = Traditional deliverable or task in the conceptual design phase.
AE&M = Availability engineering & management

Fig. 3–2 Activities to establish management policies for availability engineering and management in the project (AC1 of Figure 3-1).

- A company sends a team in search of an engineering process for reliability and maintainability. It was to be standard in all future capital projects. Once again, this is a mandate for the availability discipline.
- Companies commonly have an assortment of general policies and initiatives that call for increased asset utilization, operational integrity, total productive manufacturing and maintenance, total quality management, competitiveness, capital projects practices for life-cycle engineering and cost, computer integrated manufacturing, etc.

Each case requires that availability performance be addressed. Therefore, the final goal of this task will translate these goals, policies and initiatives those of the availability discipline in the subject project.

AC1.3. Decide the degree of availability design.   Management has a choice for the extent that availability performance needs are served in the project. The range of choices are as follows:

- To apply availability engineering and management to the goals of safety and environmental management. Thus, its application is limited to the mechanical integrity of associated critical equipment and functions.
- To apply the discipline to the plant's profit-effective availability performance.

The first extreme is the minimum requirement as management moves to comply with regulations, industry initiatives, and professional society standards. The other is an election to maximize plant income and productivity of assets. It automatically serves the goals of safety and environmental management with respect to mechanical integrity.

Does it REALLY OR is it possible
Production To maximize AND compromise safety?

AC1.4. Establish life-cycle policies for availability management. Necessary availability engineering and management policies go beyond design, construction, and startup. They include the entire life cycle.

Life-cycle policy addresses questions like the following:

- Is availability performance is to be a living design process throughout the plant's life.
- What stature, resources and structure is to be given to availability management and its functions.
- What is the business of maintenance operations? Is it to deliver profit-effective availability to perform? Or is it limited to effective and efficient repair services?

AC1.5. Establish policies for the plant in the larger production systems. Many plants are part of a larger production system. They may be one in a stream of facilities or they may be one of several plants producing the same product. Thus, management may need to establish availability policies for the plant in the context of the larger production system. These should reflect markets, resource environments, and the synergy of the plant's operations.

Does this REALLY ENCOURAGE BEING THE Best?

AC1.6. Establish consequences to be avoided.   There are always consequences that plant design and management wishes to or must avoid. These include safety, environment, productivity, financial and other strategic consequences. Management's policy for availability engineering and management must identify such cases.

Is this OVER emphasis of A.E.? CAN'T this fit other "PROGRAM s" AS well?

AC1.7. Establish availability policies for future projects and initiatives. Ideally, management policies should look beyond the subject project. An example would be a determination to make availability engineering and management standard in future capital projects and in initiatives to advance the performance of existing plants.

These policies may affect the approach taken to the subject project. For example, project deliverables may be developed as standards. Or, there may a be greater stress to widely capitalize on the project as a learning process. In any case, data development and its subsequent management should be planned to better support future projects and initiatives to advance the performance of existing plants

Criticality of Policies for Availability Engineering and Management

Management normally establishes policies for plant projects and the traditional design disciplines. This may be a formal or informal process. There may also be standing policies or a set of decisions for each project.

A conscious, formal policy process is critical to the availability discipline because it is an important addition to traditional project philosophies and approaches. The current awareness of its deliverables and methods does not equal that for the traditional disciplines. Furthermore, there will be an additional period before it is adopted as a standard. Until then, management cannot assume automatic treatment regardless of the immense performance ramifications.

TC2. Formulate Plant Operational Requirements

Definition and Purpose

Plant operational requirements translate the plant owner's needs analysis (TC1) into production and plant level requirements. It also identifies external factors that are issues and constraints affecting the requirements.

This deliverable is traditional in the conceptual design phase. Piecemeal availability performance requirements have historically a part of it.

The Scope of Operational Requirements

The statement of plant operational requirements should at least include the following:

- Operating scenarios that describe what the plant is to accomplish and how.
- Use requirements such as operations per period, operational cycles, etc.
- Regulatory requirements, industry initiatives, and design standards that affect plant design and functioning.
- Operating characteristics and constraints. These include productive capacity, output specifications, raw materials and feedstocks, utility requirements, etc.
- Operational life cycle. Concerns are anticipated life span, who will operate and maintain the plant, the characteristics of the labor force, etc. It also includes the nature and problems of plant retirement.
- External and internal environments in which the plant must operate and be maintained.
- Measures of performance such as various input/output relations, etc.
- Social, political, and regional economic issues of host governing jurisdictions; resulting desired host benefits; and requirements for slowing the evolution of host jurisdiction controls and constraints.
- Operating and business consequences management wishes to avoid.

TC3. Formulate, Evaluate, and Select from Alternate Production Processes

Definition and Purpose

There are many roads to Rome. One must be chosen. This is so for the production process. Therefore, the purpose of this deliverable is to formulate alternative production processes, evaluate them, and select the best one.

Selecting the production process is a difficult engineering problem. This is because the selection process has two dimensions. First, there are choices between processes. Second, there are choices in its subprocesses. Thus, this deliverable is to select the best alternative from many formulated possibilities.

The Process of Selecting the Production Process

The project approach to formulating, evaluating, and selecting from different production processes must incorporate the following:

- Extending the previous deliverables to study the production process requirements in each phase of the plant's life cycle. This study refines the determination of the desired and undesired outcomes and existing conditions and constraints for each of the production processes. In turn, it determines the requirements to achieve or avoid outcomes associated with conditions and constraints.
- Formulating alternate candidate production processes and associated plant concepts.
- Screening the candidates for their feasibility. This includes physical realizability, economic performance, adherence to financial and scheduling constraints, and social and political reality.
- Formulating criteria and parameters for alternate plant processes. Criteria and parameters are taken from the previously defined desired and undesired life-cycle outcomes. The criteria are ranked by relative rather than rank-order importance.

  A formal determination of criteria is critical. An omitted criterion will also be omitted from the selection of the best process.

- Selecting the best production process.

TC4. Develop the Plant Conceptual Design

Definition and Purpose

Once the best production process has been selected, the plant concept must be developed to some appropriate degree of detail.

The Process of Developing Plant Conceptual Detail

The steps for developing the plant concept deliverable are generally as follows:

- Optimize the production process.
- Optimize subsystems within the production process including subsystem configurations and equipment decisions.

- Develop the preliminary plant configuration and layout.
- Refine cost estimates as possible.

Availability Factors in the Conceptual Detail

This deliverable is the first development of hard design. There are two equivalent concepts for availability design. They are the availability (AC2) and maintenance operation (AC3) concepts described in the following sections.

The important point is that this deliverable and the two availability deliverables are iterative. Thus, they are optimized as a triad of design concepts.

The criticalness of this optimization can be seen in the top-level determinates of reliability and maintainability. They were described in Chapter 1. A significant share of the determinates were the result of hard design decisions. The remaining were the result of decisions in the design of operations and maintenance processes. Therefore, plant performance requires that hard design for reliability and maintainability be an integral part of this deliverable. In turn, they are an optimization of the factors affected by process operations and maintenance design.

Thus, this traditional deliverable provides a window of opportunity. Otherwise, the ability to achieve the maximum possible business results may be permanently lost. This will be the case if the plant concept is developed without regard for reliability and maintainability.

AC2. Develop the Availability Concept

Definition and Purpose

A rigorous conceptual development of availability performance must be added to the traditional process. It has the following purposes:

- To determine and detail the nature and value of availability performance.
- To establish the production and financial criteria and measures of availability performance.

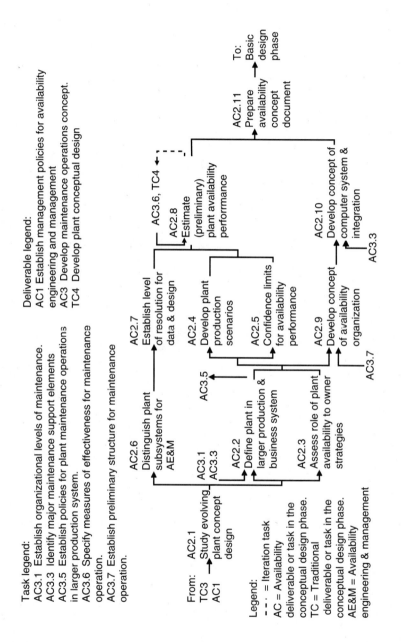

Task legend:
AC3.1 Establish organizational levels of maintenance.
AC3.3 Identify major maintenance support elements
AC3.5 Establish policies for plant maintenance operations in larger production system.
AC3.6 Specify measures of effectiveness for maintenance operation.
AC3.7 Establish preliminary structure for maintenance operation.

Deliverable legend:
AC1 Establish management policies for availability engineering and management
AC3 Develop maintenance operations concept.
TC4 Develop plant conceptual design

From:
TC3 ── Study evolving plant concept design
AC1

Legend:
-- -- = Iteration task
AC = Availability deliverable or task in the conceptual design phase.
TC = Traditional deliverable or task in the conceptual design phase.
AE&M = Availability engineering & management

AC2.1

AC2.6 Distinguish plant subsystems for AE&M

AC3.1
AC3.3

AC2.2 Define plant in larger production & business system

AC2.3 Assess role of plant availability to owner strategies

AC3.5

AC2.7 Establish level of resolution for data & design

AC2.4 Develop plant production scenarios

AC2.5 Confidence limits for availability performance

AC2.9 Develop concept of availability organization

AC3.7

AC3.6, TC4

AC2.8 Estimate (preliminary) plant availability perforrmance

AC2.10 Develop concept of computer system & integration

AC3.3

AC2.11 Prepare availability concept document

To:
Basic design phase

Fig. 3–3 Activities to develop the availability concept (AC2 of Figure 3-1)

Processes for Developing the Availability Concept

The scope of plant conceptual design varies for each plant. However, the process (Figure 3-3) should at least include the following steps:

AC2.1. Study the evolving plant conceptual design. The plant conceptual design is continually evolving. Thus, its detail should be studied carefully. The objective is to develop an in- depth understanding of the plant and the logic behind it.

AC2.2. Define the plant in its larger production and business system. The nature and value of availability performance is a function of the plant in a larger production and business system. The boundary of that system may go beyond the owner's facilities.

The larger systems include:

- Capacity, behavior, and trends in upstream or (supplier) production processes.
- Supply requirements and market characteristics faced by downstream (customer) production processes.
- Cyclical and long-term market demand trends for the plant's products. This is derived from the previous issue.
- Anticipated development phases over the plant's lifetime.
- Nature, relationship, and competitiveness of parallel existing and future facilities.
- Translation of the findings to the plant's availability performance.

AC2.3. Assess meaning of plant availability to corporate and business strategies. This task is to determine how the plant availability scheme must fit the plant owner's corporate and business-level strategies. This is important because empirical evidence shows that additional profitability and competitiveness come from owning related diversified businesses. Greater business performance is the result of the synergy between them. Thus, this fundamental issue that must be translated to the plant's availability scheme and its parameters.

AC2.4. Develop plant production scenarios. This task is to describe and quantify availability performance for various production scenarios.

It was explained in Chapter 1 that availability is an expectation of performance. As the plant operates, the probability of its expected availability gradually decreases. This is the cumulative effect of each failure in the plant. At some number of failures, availability is physically lost. The result may be lower level of productive capacity.

The scenarios are an optimal set of acceptable expected production levels. These are key production capacity levels along the curve of Figure 1-4. The figure's curve represents a scenario.

The scenarios are formed as a composite description of necessary performance. The plant should ultimately be designed for adequate expected availability at its various levels. This task allows each to be translated to availability, reliability, maintainability, and economics parameters.

AC2.5. Establish confidence limits for availability performance. An important management decision is the acceptable confidence limits of availability performance. Limits are established for each key level in the composite production scenario. The specification of these limits is the responsibility of management responsibility because they are strategically fundamental to long-term business success. It is the design team's responsibility to support the decision process.

As explained in Chapter 1, availability is a probability distribution of expected performance. It is specified as a mean availability and an acceptable upper and lower limit. These limits greatly affect plant design and its ultimate profitability even when mean availability stays the same.

AC2.6. Distinguish plant subsystems for availability design and management. The analysis and determination of availability performance has so far been at the plant level. It has also been concerned with production and business system requirements. However, it has not yet been defined at a level that can actually be designed.

Thus, the availability concept now begins to distinguish the nature, value, and top-level details of availability at the subsystem level. This is done as follows:

- Distinguish between subsystems within the plant.
- Determine the criticality of each to plant availability performance.
- Determine their interrelationships with respect to plant availability performance.
- Group the subsystems for availability analysis, design, and management.
- Establish the depth of the design and sophistication of management for each subsystem.

AC2.7. Establish level of resolution for design, data, and information Designing for availability is a substantial challenge. Therefore, it must be approached pragmatically. Otherwise, it can become an overwhelming endeavor. Therefore, the following design decisions are part of the availability concept:

Establish the level of resolution for analysis.   These decisions should include equipment, assemblies, and components. For example, plant availability performance is more sensitive to some plant items than others. Thus, the design team will want to model more critical equipment as a system of its elements. Less critical equipment may be treated as a single item in the model. There are many factors that will cause the team to make such decisions. The point is that the team must consider each plant item and subsystem for its degree of appropriate resolution.

Establish the sources and quality of the data.   This decision is connected to the level of resolution. It is concerned with the nature of sources and associated quality of the data. The decision making process must not only consider current cases, but also the possible nature and quality of data that can be developed in the future. This concern for data and information in the context of resolution may lead to a decision to invest in the acquisition and development of data and information. The revealed necessary investment may cause the team to reconsider their earlier decision (AC2.7) for the level of resolution.

AC2.8. Estimate (preliminary) plant availability performance.   The traditional conceptual design activities (TC4) involve the process of choosing and optimizing the plant scheme. This task is an integral part of those activities. The objective is to evaluate the prospective designs for expected availability performance and associated life-cycle costs.

To do this, it is necessary to develop preliminary availability and economic models. Chapters 6 and 7 describe the modeling process in the basic design phase. The principles and steps are also applicable to conceptual design-phase modeling. The difference is the sophistication of the models and possibly less fully developed data.

The steps for developing the preliminary models are as follows:

Formulate the availability model.   The initial objective is to select the software and determine the structure of model's elements. The most influential issues are the level of resolution, the confidence level of the output, the nature of the data available at the conceptual design phase, the plant design team's cycles, and the nature of the need for analysis of plant schemes, etc.

Formulate the economic model.   Once again, the initial objective is to determine the software, elements, and structure. The model is used to estimate the life-cycle costs of reliability and maintainability and, therefore, availability performance. Discussions for developing a detailed cost model are threaded throughout the tasks of the basic and detailed design phases. The principles apply here. However, the objective at this time is to be less rigorous and more concerned with understanding the cost trade-offs and comparing candidate schemes.

One dimension of the model is direct costs. They are the costs of equipment and the direct costs of returning equipment to service. The other dimension is indirect costs. They are estimated based on the factors of effectiveness (A3.6) that are established in the development of the maintenance operation concept (A3). These measures establish various resource utilization goals. Thus, they are used here to estimate cost levels associated with achieving the time to return down items to service. These time intervals are assumed in the calculation by the availability model. The cost model may reveal the need to revise them. The factors of effectiveness are also likely to be revised as the preliminary models test their reality.

It was mentioned in Chapter 2 that any economic model should be structured in the format of life-cycle income statements. The design team will surely want to expand this model to integrate production process operating cost and revenue profiles. This way, all aspects of life-cycle business performance can be explored, analyzed for optimal trade-offs and the plant scheme can be designed accordingly.

Gather data for reliability, maintainability and economic analysis. The challenge of data development (AB3) is a topic of the design tasks presented in Chapter 6. The principles, sources of data and project tasks described in the context of the basic design phase also apply here. The goal is to be practical. Management does not typically expect close confidence limit forecasts at this phase. The limits will narrow as the design progresses. At this stage the goal is to confirm that evolving plant schemes can be expected to perform to now established plant-level performance goals. The other goal is to compare different schemes or variations within them.

AC2.9. Develop a concept of the availability organization. Managing availability performance entails a well-developed organization. A large segment is concerned with maintenance operations. The remaining segments design and manage overall availability performance.

Organizational effectiveness is, therefore, important. It begins with this deliverable. The objective is to define and describe an organization as a preliminary concept. This will guide deliberations until an organization is rigorously designed in the detailed design phase (See Chapter 13).

AC2.10.   Develop a concept for computer systems and their integration.   Computer systems are the infrastructure of availability performance. This will become apparent as the deliverables of each design phase repeatedly identify requirements for defining and designing computer systems. This will be reinforced as the business processes of availability management are developed for the plant's production life.

This deliverable is to make a preliminary determination of system needs. The goal is to identify available technology and make basic selection decisions. The final product is a preliminary scheme and configuration of an integrated computer system.

AC2.11. Prepare the availability concept document.   The plant's conceptual design must ultimately documented as text and drawings. This is also true for the availability concept.

This task, then, is to prepare and document the detailed concept for achieving plant availability. Its sections should detail the findings and decisions for each of the preceding tasks. A concept for maintenance operations that correlates the concepts of availability and maintenance must also developed (AC3.9). This may be achieved in an iterative process.

The Evolving Availability Concept

The previous tasks were to explore the nature of the facility, analyze its subsystems, and develop criteria and parameters for availability performance. The resulting availability concept is not just accomplished, documented, and incorporated in overall design. The availability engineering discipline must also track changes throughout the design phases. The concept is reviewed and revised as these changes occur.

The evolution should continue throughout the plant's producing life because conditions may change for a host of reasons.

AC3. Develop the Maintenance Operation Concept

Definition and Purpose

Figure 3-1 shows that a plant maintenance operations concept is also part of a comprehensive conceptual design. It is a top-level determination of how the plant will be maintained and supported throughout its life with respect to achieving plant availability and, therefore, business performance. Thus, it sets basic policy and criteria for the design of plant maintenance operations.

The Traditional View of Maintenance Operations in Plant Design

It is tempting to regard the maintenance operation concept as relevant only to maintenance and its support elements. Design teams have historically given the issues of deliverables great importance, but little methodical attention has been them. One reason is that the teams have regarded their domain as plant design rather than maintenance.

However, the decisions made in formulating the maintenance operations concept affect the performance of equipment. Thus, the concept partially drives plant configuration and the design and selection of equipment. It is, in turn, affected by these decisions. The point is that plant design is incomplete if developed without the maintenance operation concept. Worse, it is a suboptimal one.

Process for Developing the Maintenance Operation Concept

The maintenance operation concept (Fig. 3-4) includes the following tasks.

AC3.1. Establish organizational levels of maintenance.   Organizational levels of maintenance are determined and established. These are the levels where categories of maintenance activities will be done organizationally. For example, activities to inspect or adjust certain types of equipment may be assigned to departments outside of maintenance.

There are four basic organizational levels. They are as follows:

Activities assigned to operating functions.   The guiding criteria for this choice may include the following:

- Tasks that are simple or can be accomplished within a limited time and concentration.

- Limited skill requirements for assigned maintenance personnel.

- The creation of feedback, awareness, or sensitivity to critical plant conditions for an operating group.

Activities assigned to mobile and fixed maintenance areas, functions, and personnel. This level includes all types of maintenance tasks. Different skill levels are identified within it. Maintenance facilities serve both this and the previous level.

Activities assigned to specialized owner and manufacturer maintenance organizations.   These organizations and their facilities may be remote and serve more than one plant or owner. The assigned tasks are beyond the capabilities of the previous levels.

Task legend:
AC2.2 Define plant in larger production & business
system.
AC2.8 Estimate (preliminary) plant availability
performance.
AC2.9 Develop concept of availability organization.
AC2.10 Develop concept of computer system &
integration

Deliverable legend:
AC1 Establish management policies for availability
engineering and managment.
TC3 Select the production process.

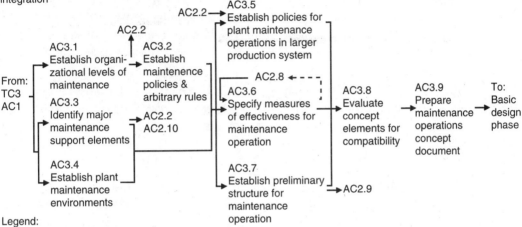

Legend:
- - ► = Iteration task
AC = "Availability" deliverable or task in the "conceptual" design phase.
TC = "Traditional" deliverable or task in the "conceptual" design phase.

Fig. 3–4 Activities to formulate the maintenance and support system
concept (AC3 of Figure 3-1)

The level may be chosen as a function of the following criteria:

- Complex or major pieces of equipment.
- Required special working conditions, equipment, methods, and
  skills.
- Special requirements for spare and repair parts and their manage-
  ment.
- Special environmental and other control requirements.
- Economies of scale and specialization.

Activities assigned to contract maintenance organizations. This
choice cuts across the previous organizational levels. Besides personnel
cost flexibility, the choice may be driven by the following:

- Contract services that are better organized, skilled, or developed than
  those available internally.
- Existing resources that cannot serve the short-term work-load
  requirements and still maintain the quality of its normal work load.

AC3.2. Establish arbitrary maintenance rules.   Every plant and its owner has certain rules and policies that maintenance planning must observe. Some are arbitrary; others are established by analysis. Examples are:

Management may make arbitrary decisions for the maintenance response to certain types of failures.   There may be categories of equipment subject to such rules. An example is to immediately repair a failed item in a pair of redundant items. This deliverable is to search out such cases for incorporation in the ultimate maintenance operation scheme.

The point is that such rules may be widely accepted in an organization. The plant culture may even consider them sacred. Plant management may even feel insecure without them. Both situations may be the case regardless of actual merit. Hopefully, some will be eliminated over time as the results of availability design indicate a need for change.

Basic policy may be established for equipment, assemblies, and components to be replaced rather than repaired.   Choices are made along a continuum. At one extreme is full replacement; at the other end is full repair. The middle position is a mix of both.

Basic testing policies may be established for groups of equipment, assemblies and components.   These policies are especially relevant to electronic equipment. The choices for testing are as follows:

• Testing versus no testing.
• Manual versus automatic.
• Internal versus external.

Policy may be established for the number of manufacturers to supply widely used plant equipment.   Pumps are an example of such equipment. The policy affects the degree to which the benefits of standardization and interchangeability are possible. It may also affect possibilities for partnering relationships with equipment suppliers. A large number reduces these opportunities. A small number reduces the opportunities for optimization in each application.

AC3.3. Identify major maintenance support elements.   Major maintenance support elements must be identified. These include testing and equipment, support facilities, skill categories, major spare parts, computer systems and software, management functions, etc.

AC3.4. Establish plant maintenance environments. Various plant maintenance environments exist naturally or are created by design. Special conditions may be required for some maintenance activities. Alternately, plant design may create difficult maintenance conditions. Thus, the purpose of this deliverable is to identify them. It is then make plant design decisions or accept the conditions as a given. Both are ultimately reflected in the design of maintenance operations.

AC3.5. Establish policies for plant maintenance operations in the larger production system. Few production facilities exist in isolation. This is so even when remotely located. The plant may be required to share common resources and functions. It may also be required to apply standard methods and software. Thus, it is necessary to establish when the maintenance operations concept must be designed to recognize such cases. This assessment includes existing and future production plants and organizational entities.

AC3.6. Specify measures of effectiveness for maintenance operations. Measures of maintenance operations effectiveness must be established. Examples are maintenance-hours-per-production- hour, staff utilization rate, parts supply and equipment responsiveness, testing and handling equipment reliability, support facility utilization rate, maintenance staff levels, administrative response times, etc. These must be correlated with the availability and economic performance estimates of AC2.8.

AC3.7. Establish a preliminary organizational concept for maintenance operations. A major factor in availability performance is organizational effectiveness. Thus, a preliminary structure for all maintenance operations responsibilities, authority, and organization must be developed. This concept should be integrated with the organization's concept for overall availability management (AC2.9). Thus, it becomes a subset of that concept.

The preliminary formulation should account for existing local and overall organizational structures. Thus, this task is to understand the purpose, advantages, and disadvantages of existing cases.

AC3.8. Evaluate the concept elements for compatibility. Because the above choices are related, it is necessary to formulate them as a system of compatible policies. Therefore, the above concept decisions and details are evaluated for compatibility with the previously described deliverables.

AC3.9. Prepare the maintenance operation concept document. The final concept should be prepared and documented. It should be developed along with its equivalent for the plant design and availability concepts (AC2.11).

The Evolving Maintenance Operations Concept

The maintenance operations concept does not take its final form as a one-time project deliverable. Like the availability concept, it evolves as the plant design progresses through its phases.

This evolution continues throughout the plant's life. It is a part of the living design as operational requirements are changed and improvement initiatives occur.

Bibliography

Blanchard, Benjamin S. *Logistics Engineering and Management*. 4th ed. Englewood Cliffs, N. J., Prentice Hall, 1991.

Bloch, Hienz P. *Practical Machinery Management for Process Plants*. Volume 1. Improving Machinery Reliability. 2nd ed. Houston, Tx., Gulf Publishing Company, 1988.

Bloch, Hienz P. and Geitner, Fred K. *Practical Machinery Management for Process Plants*. Volume 3. Machinery Component Maintenance and Repair. 2nd ed. Houston, Tx., Gulf Publishing Company, 1990.

Center for Chemical Process Safety. Guidelines for Technical Management of Chemical Process Safety. American Institute of Chemical Engineers, New York, 1989.

Hofer, Charles W. and Schendel, Dan. *Strategy Formulation: Analytical Concepts*. St. Paul, Minn., West Publishing Company, 1978

Ostrofsky, Benjamin. *Design, Planning and Development Methodology*. Englewood Cliffs, N. J., Prentice Hall, 1977.

Robock, Stephen H. and Simmonds, Kenneth. *International Business and Multinational Enterprises*. 4th ed. Homewood, Ill., Richard D. Irwin, Inc., 1989.

Stebbing, Lionel. *Quality Assurance, the Route to Efficiency and Competitiveness*. Chichester, West Sussex, Ellis Horwood Limited, England, 1993.

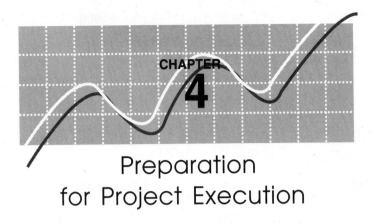

Preparation
for Project Execution

Introduction

It is critical to ultimate plant availability performance to include certain deliverables in the conceptual design phase. These deliverables were the subject of Chapter 3.

There is another crucial requirement. It is careful planning for the availability discipline in the project. Without it, design for availability performance will lack critical mass in the design phases. Its issues will still be regarded as important, but treated piecemeal if at all.

The availability engineering and management processes of this book are equally applicable to existing plants. However, this chapter is written for the plant capital project. It discusses the preparation for the availability discipline in the plant or capital project design phases.

The chapter is written with the assumption that the project will at some time be assigned to a contract engineering and construction organization. However, the requirements are still generally applicable if this is not the case.

The described project preparation deliverables and associated tasks are universal. Their exact approach and terminology are unique to each owner, project contractor, and project.

It may be best to regard the deliverables of this chapter as what "has to happen" as part of the owner's project planning and control responsibilities. They resulted from the repeated observation of what has failed to happen in past projects.

The Challenges of a New Discipline

The owner's project management team has a fundamental challenge: How to get the project contractor to meet their needs for achieving profit-effective plant availability performance. This, of course, is preceded by the need for the owner and project manager to know what they need. That requirement is fulfilled by this book.

The experience of many project engineers has shown that moving the project contractor to respond is not usually accomplished by making a simple request. One reason is the current position of the availability discipline in industry. It creates many pitfalls to avoid when working to assure that the need is met.

The Position of the Availability Discipline in Industry

When planning a project, it is important to recognize the position of the discipline in the owner's industry. As this book is published, the technology of availability analysis methods and design tools is widely available. However, it is usually applied reactively and piecemeal to existing plants. It does not often appear in capital projects. When it is applied, it is usually without the structure of the availability discipline.

Some owners and project contractors have long known that the elements of availability are important to business results. They have attempted to treat them from within the traditional disciplines. The result has been frustration and defeat for the lack of an appropriate engineering and management discipline.

Therefore, the availability discipline is a new experience to the many participants in the traditional project approach. The inclusion of the discipline is also more of a project management challenge rather than one of inventing new design methods for two reasons. First, as mentioned, its individual methods already exist. Second, by virtue of this book, there is a discipline to apply them in a system of project deliverables.

Fortunately, owners and their project contractors are astute in project management practices and control systems. Thus, they are quite capable of integrating and managing availability design along with the traditional disciplines.

Pitfalls to Avoid

The above described position of the availability discipline in industry creates specific pitfalls to be avoided by affirmative preparation for project execution. They are as follows:

- Loss of choice between competing project contractors in the process of garnering an appropriate response to availability requirements.
- Disruption in the project progress when it is later necessary to remedy project execution plans by incorporating a design for availability requirements.

Owners usually prefer that there are several legitimate project contractors competing for a project. A new design requirement can frustrate that goal. This is so when not all of the competitors respond to them. Reasons for this may be the following:

- The contractors' history or lack of specialized capabilities for meeting the new requirement.
- The contractors' cultures result in minimal response to new requirements.
- The owner's explicit but poorly stated requirements. Management knows what it wants, but does not know how to ask for it. A solution to this problem is the objective of this chapter.

Owners and contractors wish to avoid disruptive changes in the project's work scope, tasks, and costs. This can easily be the case for a new project design requirement. As the project progresses, any deficiencies will be revealed in the contractor's initial response to availability requirements.

There are two options when this is the case. One is to leave the need untreated. The other is to take remedial action. That action may be disruptive. It will at the least result in project chaos and budget overruns. It may also result in disrupted schedules and dented careers.

Summary of Requirements

There are three fundamental requirements in preparation to design the plant for availability performance. They are:

- The development of availability elements in the documents for screening, selecting, and contracting a project contractor (AP1).

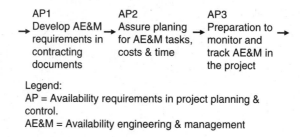

Legend:
AP = Availability requirements in project planning & control.
AE&M = Availability engineering & management

Fig. 4–1 Activities to prepare for availability design in project execution.

- Appropriate planning of project tasks, cost, time, and resources for availability design (AP2).
- Preparation to monitor and track the project contractor's management of the availability discipline (AP3).

In Figure 4-1, these requirements are shown as a flowchart of deliverables. Subsequent flowcharts show their respective tasks. They are coded in the charts as "AP" to denote *availability* ("A") requirements for *project planning* ("P") and control.

AP1. Develop Availability Requirements for Contractor Screening, Selection, and Contract Documents

Definition and Purpose

The first deliverable is to develop availability discipline requirements in the traditional screening, selection, and contractual documents. The objectives are to:

- Define the project-specific approach to availability engineering and management.
- Document the associated general policies and practices associated with the approach.
- Develop the owner's screening, proposal request, and contract documents for availability design in the project.

Deliverable legend:
AP2 Assure planing for AE&M tasks, costs, & time

AP1.1	AP1.2	AP1.3	AP1.4	
→ Form project →	Prepare document →	Assess project →	Write sections →	To:
approach to	of project policy &	contracting	of contracting	AP2
availability	approach to	documents	documents	
design	availability design			

Legend:
AP = Availability requirements in project planning & control.
AE&M = Availability engineering & management

Fig. 4–2 Activities to include availability engineering management requirements in the documents to select a project contractor (AP1 of Figure 4-1).

Process to Develop Project Contractor Selection and Contract Documents

The deliverable to develop the contract documents to select the project contractor is flowcharted in Figure 4-2. It entails the following tasks.

AP1.1. Formulate the project-specific approach to availability design. At this point the basic approach to availability engineering and management in the project will have been developed. This task is to chart its basic path through the project. This is done with a flowchart that describes the project-specific process for availability design.

The life-cycle process described by this book is a baseline approach. Therefore, the task of this deliverable is to review these baseline processes and adjust them accordingly.

AP1.2. Prepare a document of general policies and practices for availability design. Principles of quality assurance stress the importance of this task, which is to formally declare and document management's intent and approach to the design of availability performance. In essence, the documents are the general equivalent of a quality manual.

The document should include the following:

- Basic policies to guide the project team to diligently achieve plant design and management for availability performance. These policies should draw from the conceptual design-phase deliverable that established associated management policies (AC1 in Chapter 3). They may be adopted directly or adjusted to specific project management goals.

- Convert the flowcharted approach to a basic procedure of practices. The objective is to prepare a summary-level list of flowcharted deliverables and their basic tasks and purposes. The details will be developed in later project plans and controls.

AP1.3. Assess project contracting documents. The next task is to assess the series of documents leading to the final contract for project services with respect to availability design. It is to determine what has been included, omitted, or needs to be enhanced.

The contracting documents are used to screen, evaluate, select, and commit the project contractor. They also define how the progress and quality of the project will be confirmed. They are immensely important to the entire project because they contractually formalize the owner's policy and approach to availability performance design.

All that follows will reflect the quality of their content. The availability discipline can be lost to plant design. Substantial money and time may be required to remedy the situation. Ideally, the documents should insure that planning details are developed in a timely fashion.

AP1.4. Write sections, clauses, and terminology for contracting documents. The sections, clauses, and terminology for availability engineering and management are written into the traditional enquiry, selection, and contract documents. That is the purpose of this task. The previous task determined the specific needs.

These documents should be formulated according to the three criteria relative to traditional documents. They are:

- To fit the nature and format of the traditional documents.
- To be more comprehensive.
- To provide more explicit and extensive text.

As mentioned earlier, a challenge to project management is the historically limited exposure of the plant and contractor personnel to the availability discipline. Thus, there is not a well-traveled trail to follow as there is in the traditional disciplines. The last two criteria reflect this difference.

AP2. Assure Planning for Availability Tasks, Costs, Time, and Resources

Definitions and Purpose

The second deliverable is to overview, support, and sometimes guide project contractors as they develop project plans for the availability discipline. The discipline details are only a part of the overall plan of interdisciplinary project tasks, resources, costs, and schedules.

Requirements for this deliverable may arise at two stages: when contractors are competing for the project and during the selected contractor's project mobilization activities.

Accordingly, the objectives of the deliverable are as follows:

- To clarify each candidate contractor's understanding of the owner's requirements and approach for the design of plant availability performance.
- To subsequently verify that these are included in the contractor's project scope, tasks, schedule, cost, and control systems.

Process to Assure Development of Project Plans for Availability

This deliverable (Figure 4-3) entails the tasks described in the following sections.

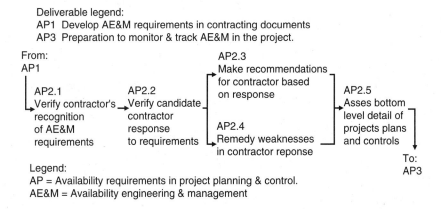

Deliverable legend:
AP1 Develop AE&M requirements in contracting documents
AP3 Preparation to monitor & track AE&M in the project.

Legend:
AP = Availability requirements in project planning & control.
AE&M = Availability engineering & management

Fig. 4–3 Activities to support the contractor's development of project detail as a candidate or selected contractor (AP2 of Figure 4-1).

AP2.1. Verify and refine the contractors' recognition of requirements.   This task is to verify that each candidate project contractor understands the following:

- The owner's commitment to the design for plant availability perfor-mance.
- The value of plant availability performance to the owner.
- The owner's general policy and determined necessary approach to achieving that performance.

AP2.2. Verify contractor response to requirements.   This task is to verify that each received proposal has responded fully to the owner's stated avail-ability design requirements.

AP2.3. Recommend the project contractor based on the response.   The previous deliverable assessed the response of each candidate to availability and associated safety management requirements.   Accordingly, the assess-ment team should make recommendations for the selection of a contractor. This recommendation should be made only with respect to availability engi-neering and management requirements.

AP2.4. Remedy weaknesses in project contractor responses.   This task is optional. It reflects the owner's philosophy for competition, project award, and contract negotiation.

Availability engineering and management is a new discipline. The owner may not be well served by eliminating contractors based on their ini-tial response to stated requirements. Instead, it may be better to work with the short-listed or the selected project contractor to rectify shortcomings in their response to availability design requirements instead of eliminating the contractor from competition.

AP2.5. Assess bottom-level detail of the project plan and controls.   At some time during the project, the project contractor will prepare details for the bottom-level availability engineering tasks to be incorporated in the project plan and control systems. This task is to assess the quality of these details. It should determine if the details were derived by identifying the fol-lowing:

- The traditional project disciplines.
- The top-level project deliverables including various plant layouts, engi-neering diagrams, subsystem designs, procurement stages, project reviews, etc.

- The bottom-level tasks of each discipline with respect to each availability deliverable.
- The necessary bottom-level availability engineering tasks required in response to each task of the traditional deliverables.
- The bottom-level tasks required of the traditional disciplines in response to those of the availability discipline.

This process is not a unique determination. In fact, it occurs in the traditional project design and management disciplines. Owners and project contractors long ago captured the findings in detailed project flowcharts, procedures, and standards. Thus, this determination extends historical detail to include availability engineering and management.

AP3. Prepare to Monitor and Track Availability Engineering

Definition and Purpose

The third deliverable recognizes the need to prepare for monitoring and tracking the project contractor's management of availability engineering. Thus, its objectives are as follows:

- To assure the efficient and effective execution of availability engineering as a discipline in the project. *Efficient* means to be in-budget and on-time. *Effective* means that the tasks are completed in good form.
- To leave a trail of project procedures for future capital projects.
- To capture and capitalize on findings and insights pertinent to future capital projects and initiatives to advance the performance of existing plants.

The last two objectives recognize the natural benefits of adding another dimension to a traditional process. Properly managed, these benefits go far beyond the immediate capital project.

Process for Preparation to Track and Monitor Availability Design

The tasks of this deliverable (Figure 4-4) are presented in the following sections.

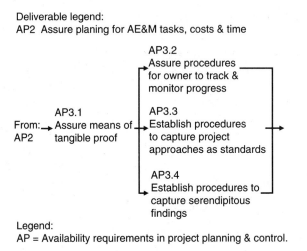

Deliverable legend:
AP2 Assure planing for AE&M tasks, costs & time

Legend:
AP = Availability requirements in project planning & control.

Fig. 4–4 Activities to monitor and track the availability engineering process (AP3 of Figure 4-1).

AP3.1. Establish a means of tangible proof. This task is to develop a means by which the owner will have tangible proof of project performance. This is to assure that the discipline's design tasks are progressing in an efficient, effective, and integrated manner.

This is more than just a principle of quality management. As mentioned, availability is a new design and management discipline for plant design projects and is likely to be so until sometime in the future.

This suggests the need for a rigorous approach to providing tangible assurance. The owner will expect the project contractor to demonstrate more diligent than normal project planning, control, and management. This is compared to the control approaches for the deliverables of traditional disciplines. This task, then, is to develop such a system.

AP3.2. Establish procedures for the owner to track and monitor progress. This task is to assure that there are procedures for the owner to monitor and track the progress of availability design and management tasks.

There are three dimensions of project management to be monitored, tracked, and evaluated. They are as follows:

Disciplines management: This is concerned with functions of acquiring, developing, and assigning personnel to design tasks.

Technical review and support: This includes functions to assure that availability engineering tasks are being done correctly and functions to provide technical guidance to design personnel charged with the tasks.

Discipline lead: The people who perform these functions are often called the "people in the middle." This is because these positions have two responsibilities. First, to satisfy project management in terms of in-budget and on-time task execution. And second, to satisfy discipline management in terms of task quality and resource utilization.

AP3.3. Establish procedures to make project approaches a standard. A new discipline in the traditional project will create new trails. Once created, they may become owner and contractor standards. Many of the deliverables discussed in previous sections and their tasks are good candidates for becoming operating procedures.

Therefore, this task is to establish procedures to standardize the following:

- Project plans, controls, and procedures.
- Associated availability engineering project procedures.

AP3.4. Establish procedures to capture serendipitous findings. This task is to formulate procedures to identify and capitalize on serendipitous findings and results. This requirement recognizes the following three possible beneficiaries:

- Soon-to-begin management of the subject plant.
- Other existing production facilities.
- Future production facilities.

Capitalizing on these discoveries requires procedures for the following:

- Communicating possibilities to the appropriate owner and contractor personnel.
- Formally establishing if there is an interest in developing these possibilities in the subject project or the above other applications.

Summary

The deliverables discussed in this chapter are project management practices applicable to contracting and planning. If these project management requirements are not addressed, the project will most likely fail with respect to availability engineering and management. The consequences over the plant's lifetime may be hundreds of millions of dollars in lost income and therefore, a substantial long-term reduction in the return on the owner's assets.

Bibliography

Center for Chemical Process Safety. Guidelines for Technical Management of Chemical Process Safety. American Institute of Chemical Engineers, New York, 1989.

Ostrofsky, Benjamin. *Design, Planning and Development Methodology.* Englewood, N.J., Prentice Hall, 1977.

Stebbing, Lionel. *Quality Assurance: The Route to Efficiency and Competitiveness.* Chichester, West Sussex, England, Ellis Horwood Limited, 1993.

PART THREE

Basic Design Phase

CHAPTER 5

Foundations to Maintain Integrity of the Availability Scheme

Introduction

The previous section described availability engineering and management activities in the conceptual design phase. This and the next four chapters describe the discipline, its deliverables and their tasks as part of the basic design phase. A final chapter in this section briefly describes the traditional phase activities.

Basic Design Phase Defined

The basic or preliminary design phase has traditionally accomplished the following:

- The optimization of subsystems with respect to special aspects left open by the conceptual design phase.
- The development of various engineering diagrams.
- The sizing and selection of equipment.
- The start of the procurement of long-lead-time equipment.

What Is and Is Not Accomplished

It is important to recognize what is and what is not achieved by the traditional design activities. A process is selected and developed. A plant scheme able to achieve its production process parameters is defined. The design team has intuitive expectations that the scheme can perform for continuous extended periods of time.

However, the traditional conceptual and basic design phases do not result in engineered expectations for continuous mechanical ability to perform, nor do they determine the financial ramifications of achieving them.

This is because traditional plant design is centered on the production process. Process design searches for an optimal design of inputs and outputs with minimal generation of wastes. The equipment is primarily selected for its ability to perform within the production process parameters in the most cost-effective manner. This is opposed to an equipment selection process that also takes into account a designed set of reliability, maintainability, and associated economic parameters. Therefore, the traditional design process is best equipped to optimize the plant production process. However, it is not effective for optimizing plant availability performance. The result is a plant design that is marginal compared to the performance that is possible.

Availability Engineering in the Basic Design Phase

Scope of accomplishment. Availability engineering in the basic design phase is focused on the modeling, analyses, and design of performance. Accordingly, the activities shown in Figure 5-1 are designed to accomplish the following:

- Incorporate availability requirements in equipment design and specifications.
- Develop an infrastructure for change, improvement, and data management with respect to life-cycle availability performance.
- Analyze the plant design for potential failures and determine appropriate maintenance strategies for each.
- Verify that the plant can perform in accordance with the plant owner's specified availability.
- Determine the most profitable scheme for achieving that specified performance.

Evolving design. The design for availability performance evolves as overall plant design progresses. Its engineering processes begin along with the traditional design events and are then refined as plant detail increases.

Thus, plant design must be continually assessed for reliability, maintainability, and associated economics. This provides the feedback on which design changes are made to achieve specified availability performance. The timing is important because it becomes increasing costly and less practical to make changes as the project progresses.

Window of opportunity for the full optimization of availability performance. Availability engineering initially affects the plant's hard design because many of the top-level factors (see Chapter 1) of reliability and maintainability are associated with equipment configuration, sizing, and selection. Once solidified, only the reliability and maintainability factors associated with maintenance operations or affected by production unit operations are still highly flexible.

The goal of a proactive basic design phase is to optimize the full set of factors for maximum plant business performance. This suggests that there is a window of opportunity. Therefore, the goal is to capitalize on the opportunity to maximally optimize plant availability performance.

It is possible that availability engineering may not have been practiced from the very beginning of the project. The factors associated with plant hard design may no longer be flexible. Consequently, those related to maintenance are aligned to complement plant hard design and the intended process operation. Therefore, optimization is now generally limited to these factors.

Organization of the Discussion

The basic design phase is presented in six chapters. Thus, the phase deliverables are grouped as follows:

- Infrastructure for maintaining and advancing the integrity of the availability scheme (Chapter 5).
- Detailed development of the plant as a system of reliability, maintainability, and economic performance parameters (Chapter 6).
- Development of availability and financial performance models (Chapter 7).
- Analysis of potential failures and the development of maintenance strategies (Chapter 8).
- Analysis and optimization of plant availability performance (Chapter 9).
- Availability-centered review of the traditional basic design phase deliverables (Chapter 10).

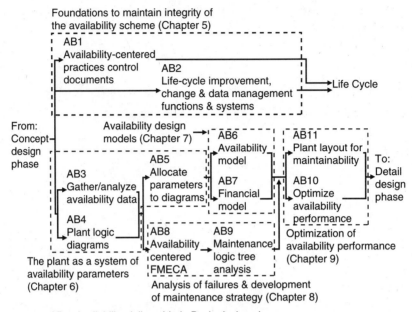

Fig. 5–1 Availability engineering deliverables for the basic design phase.

The basic design phase availability engineering deliverables are represented in Figure 5-1. They are coded as "AB" to denote an *availability* deliverable in the *basic* design phase.

The associated management chapters are bounded in the figure. Accordingly, the figure shows that the following two deliverables are the subject of this chapter.

- The development of availability-centered practices control documents (AB1).
- The development of the functions and systems to manage, change, and improve data (AB2) associated with the availability scheme.

AB1. Develop Availability-Centered Practices Control Documents

Definition and Purpose

This deliverable develops the documents that serve the need for consistent availability engineering and management practices. Individuals and teams need guidance to assure that availability issues are properly treated in their work practices. This is supported by developing availability-centered standards, checklists, and procedures. These are called *availability-centered practices documents.*

Availability-Centered Practices for the Plant Life Cycle

There is a tendency to think of availability-centered practices with respect only to plant design. This is especially true for capital projects. Ideally, they are developed to support availability engineering and management throughout the plant's life.

Such development should include the following:

- Project planning and management.
- Identification and numbering systems for plant elements.
- Availability issues for traditional disciplines.
- Reliable equipment specifications.
- Equipment management practices during construction.
- Availability requirements in the plant startup stages.
- Operating procedures that affect plant availability.
- Auditing, testing, and evaluation of plant functioning.
- The development of maintenance operations.

Project planning and management. Practices documents may be developed for project planning and management. They should include the following objectives.

- Assurance that design for availability performance is included in project management procedures and controls.
- Assurance that project task planning and costing are inclusive with respect to availability design.

Identification and numbering systems.   The master equipment list and the plant numbering, labeling, and color coding systems are part of traditional design. However, they are fundamentally important to life-cycle availability engineering and management for the following reasons:

- They are critical to the analysis and design of availability management and maintenance operations because of the vast amount of interrelated detail. Therefore, the organizational effectiveness of availability management is significantly dependent upon the identification and numbering system.
- They enable maintenance operations to better and more readily identify a subject plant item.
- They are fundamental to the efficient and accurate collection of availability data and information.
- They provide a basis for directing personnel to observe special procedures and precautions.

Availability issues for traditional disciplines.   Chapter 1 introduced the top-level factors of reliability and maintainability. Approximately half of them are associated with plant design decisions. Thus, availability-centered documents translate the factors to the design deliberations of each discipline.

This may be a special challenge for some disciplines. Their engineers may assume that availability design is only minimally relevant to their discipline. The reliability and maintainability factors are used to test that assumption as follows:

- How could each item to be designed affect each factor?
- What are the specific design points associated with those affects?

Reliability in equipment specifications.   Closely related to the previous practices are the traditional performance and purchase specifications. The concern is whether they include reliability, maintainability, and associated economic parameters.

Another major concern is the data and recommendations to be provided by the supplier. These include:

- Maintenance task detail and procedures.
- Equipment specification sheets.
- Cross-section drawings showing equipment assembly.
- Bill of materials.
- Recommended provisioning of spare and repair parts and components.
- Test data.

Equipment management practices during construction.   An important issue of quality control is how the integrity of a produced item will be maintained downstream. This includes the construction phase. Some equipment may be sensitive to handling, staging, and warehousing practices. Therefore, availability-centered practice documents are prepared for equipment whose reliability and maintainability are sensitive to such practices.

Availability requirements in the plant startup stages.   Section 5 describes the considerable involvement of availability engineering and management in plant startup. It follows that there are considerable possibilities for applicable availability-centered practice control documents.

Operating procedures that affect availability.   Plant production process operations determine operating conditions. Their consequences are manifested in the reliability component of the availability equation (Equation 1-1).

Availability analysis will discover where long-term availability performance can be affected by periodic operating practices. The long-term productive capacity is reduced beyond the value of very short-term results. Thus, availability-centered documents may take the form of agreements to operate in accordance with certain practices and constraints.

Auditing, testing, and evaluation of plant functioning.   Ideally, all aspects of plant functioning are periodically subjected to auditing, testing, and evaluation. Availability-centered practices must be developed for the following:

- Functions directly subject to assessment by availability management functions.
- Functions subject to assessment by other management functions. An example is production process safety management, which must evaluate availability practices in their assessments of mechanical integrity.

Development of maintenance operations. Practices documents can be developed for the design of maintenance support functions, facilities, equipment, and training. These will naturally evolve as the availability scheme is developed by the many deliverables and tasks that are the subject of this book.

These practices are relevant once the plant is operational because they provide checklists for guiding periodic initiatives to realign maintenance operations. This occurs as operating conditions change over time.

Timing in the Development of Availability-Centered Practice Control Documents

The development of availability-centered practices is described here as part of the capital project. Once developed, some will become baseline practices for future projects. If such a baseline exists, it should be reviewed and adjusted to fit the subject plant. If it does not exist, this task is to develop them for the project.

Availability-centered practices should be developed as early as possible in the project because of their effect on the other project design disciplines. Furthermore, maximum affect may be a short-lived opportunity.

The development of life-cycle practices is obviously a large task. Since time is of the essence, they must to be approached in stages of priority. The stages are generally as follows:

- Practices associated with project management, design, specifications, and procurement.
- Practices for startup, operating, and maintenance procedures and systems.
- Practices for auditing, testing, and evaluating practices throughout the plant's commercial production life.

Preparation is expedited by drawing upon published sources. Publications are available for types of equipment, engineering disciplines such as human factors and reliability, details for plant functions and activities, availability-centered checklists, etc. Eventually, future projects will draw on the company's own set of baseline practices.

Tasks to Accomplish the Deliverable for Availability-Centered Practices Control Documents

As previously mentioned, developing availability-centered practice documents can be a huge endeavor. It must be approached carefully. The tasks for accomplishing the deliverable are described in the following sections and presented as a flowchart in Figure 5-2. The flowchart includes sequences and iterations with tasks in other deliverables.

AB1.1. Identify the organizational functions pertinent to each phase of the plant's life cycle.   The organizational functions in each phase of the plant's life cycle must be identified for the list to be comprehensive. The list should not be limited to operational functions. The functions must also be project related. Thus, the definition of functions includes project management, control, and design disciplines.

Task legend:
AB2.2 Develop availability element & linkage system.
AB2.5 Define the data management system.
AB2.6 Define interim and permanent structure, authority, and responsibility.
AB6.1 Analyze model uses over plant life.

AB8.6 Identify the consequences of each failure.
AB8.8 Determine organizational solutions to failures.
AB11.2 Assess plant layout against layout-critical points.

Deliverable legend:
TD5 Expediting and quality assurance & control.

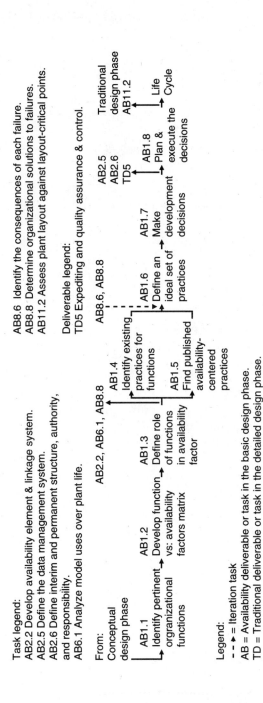

Legend:
- - ▶ = Iteration task

AB = Availability deliverable or task in the basic design phase.
TD = Traditional deliverable or task in the detailed design phase.

Fig. 5–2 Tasks of develop availability-centered control documents (AB1 of Figure 5-1)

AB1.2. Develop a function versus a reliability and maintainability factors matrix. The next task is to identify which functions affect plant availability performance. This is done by forming a matrix. Its axes and contents are defined as follows:

- The functions and their phases are positioned along one axis.
- The top-level factors of reliability and maintainability are positioned along the other axis. These factors were introduced in Chapter 1.
- How each function affects the associated factor is identified at the intersection.

AB1.3. Define the role of each function in the subject factor. The previous task determined when a function is relevant to a reliability or maintainability factor. This task is to define that relevance. This includes:

- How the function directly and indirectly affects each associated factor.
- When the consequence occurs.
- Relative anecdotal importance of the identified effects.

Figure 5-2 shows that the findings flow to tasks in other deliverables. They are as follows:

- A system is developed to trace the elements of availability management and their linkages (AB2.2, Chapter 5). It is part of the deliverable to develop life-cycle improvement, change, and data management functions and systems. The matrix provides initial information.
- A financial model is developed by a later deliverable. A task is required to evaluate its use in plant and business management (AB6.1, Chapter 7). The results of this task may help resolve that issue.
- The failure modes, effects, and criticality analysis (AB8, Chapter 8) will ultimately determine organizational solutions for each significant failure. The choices are design, production operations, maintenance functions, and quality assurance. The task to make those choices (AB8.8) will draw upon the matrix in the search for solutions.

AB1.4. Identify existing practices documents for each function. Some of the functions may already have practices documents. These should be identified and evaluated. The issues are their content, quality, and relevance to availability engineering and management.

AB1.5. Search out availability-centered practices provided by literature. Considerable literature (i.e., Blanchard 1991 or Pack, et al.,

1985) is available to help formulate reliability, maintainability, and availability practices. The objective of this task is to identify, gather, and evaluate them.

AB1.6. Define an ideal set of availability-centered practices control documents. An ideal set of availability-centered practices documents must be defined. Developing a comprehensive set of practices is an immense task. Thus, they must be developed selectively over time. The ideal formulated by this task provides a benchmark for that determination.

Figure 5-2 shows that this description may iterate with the later FMECA deliverable. The specific tasks are as follows:

- Failures identified to have significant consequence will confirm the need to develop specific practices (AB8.6, Chapter 8).
- A basic organizational solution is selected for each failure (AB8.8). The selection process confirms the need for specific practices.

AB1.7. Make development decisions. The next requirement is to make decisions to develop practices. This is done with the previously defined ideal as a baseline. The decisions to be made are the following:

- The practices to be developed.
- When in the plant's life each practice is to begin.
- How each practice will be packaged. Options include inclusion in existing procedures, checklists, etc. The alternate extreme is to write them as separate documents.
- The functions and expertise necessary for their development, use, and enforcement.

Figure 5-2 shows that the task results flow to the following tasks in other deliverables:

- The practices are included in and managed by the data management system. Thus, the decisions of this task are incorporated in the definition of the overall system (AB2.5, Chapter 5).
- An organizational scheme is developed for the improvement, change, and data management organization (AB2.6). The decisions of this task flow to that development. Thus, they are integrated into the overall scheme.
- The results identify requirements to the traditional deliverable for quality assurance (TD5, Chapter 14). This is because some practices are selected for development according to important quality assurance requirements.

AB1.8. Plan and execute the development decisions. This task is to formulate a plan to develop availability-centered practices. Its scope should include immediate and longer-term development.

Figure 5-2 shows that the resulting practices documents will immediately pass into service. This occurs when:

- They are applied to traditional design deliverables (Chapter 10). The objective is to assure that reliability and maintainability design issues are included in the traditional project tasks.
- They are applied to the availability deliverable for assessing the plant layout for maintainability (AB11, Chapter 9). The affected task is the assessment of layout-critical points of the availability scheme (AB11.2).

AB2. Develop Life-Cycle Improvement, Change and Data Management Functions and Systems

Definition and Purpose

Figure 5-3 shows that there are four interrelated management functions and systems for maintaining and advancing availability performance. They are:

Improvement management: The continuous testing and evaluation of plant performance that begins in the design phases and continues throughout the plant's life. This includes failure analysis and data trending.

Change management: The response to performance issues. Change has origins beyond the quest to improve plant availability performance. Change will be discussed in an upcoming section.

The baseline design: The benchmark against which improvements and change are identified, evaluated, and developed.

Data management to serve the first three functions.

Improvement Management

The continual improvement of availability and its economics is the fundamental goal of life-cycle availability engineering and management. This quest begins with the design phases. The first goal is to formulate a plant

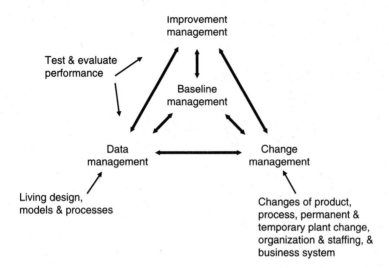

Fig. 5–3 Integration of improvement, change and data management.

design that will surpass the profitable availability performance of its predecessors. The other is to improve that design as it evolves.

The vision of improvement should continue throughout the plant's life. Availability management systems and functions must continually test and evaluate the plant in search of opportunities. This is necessary because what is known about the plant increases over time. Furthermore, the initial design is merely a set of assumptions and calculations.

Change Management

Change management is closely associated with the management of improvement. Its goal is to assure that plant changes do not unknowingly undermine its availability scheme, and that positive local changes do not have a negative net effect at the plant level.

Regardless of its intent, change can cause the creeping reduction of availability performance because very subtle hidden pitfalls may be created. They will ultimately make themselves known through a visible, possibly major, setback.

Sources of change.   Plant change that can affect the availability scheme may come from the following sources:

- Process changes: These occur as the plant attempts to maintain the production process in the face of unforeseen events. They may be temporary or permanent. Examples are compensating for equipment unavailability, improving the production process, producing a new product, or product grade, changing the production rate, etc.

- Permanent plant changes: These are necessary for a host of reasons. One is to incorporate new production equipment technology. Another is to increase the plant's productive capacity by eliminating constraints.
- Temporary plant changes: These temporarily change the plant configuration or equipment in some way. Temporary changes require close scrutiny because the controls that exist for initial design and permanent change may not exist or be easily mobilized.
- Organizational change: Organizational change has two dimensions. The first is the structure of functions responsibilities and authorities. The other is the staffing level. Both can easily undermine the availability scheme.
- Market changes: Changing market conditions can lead to any of the previous conditions. An expanding market creates the need to squeeze greater productive capacity from the existing plant. Eventually, it is necessary to remove constraints in the production process. A cyclically shrinking market dictates far-reaching plant change. The goal is to reduce productive capacity to an appropriate level. Availability performance is responsive to both changes of direction.

Consequences of change.   Changes may have small or large consequences for the availability scheme. They may have the following effects:

- They may create new operating conditions. The expected reliability of affected plant items may change significantly. In this case, the expectations for availability performance must be changed.
- Changes in plant management structure may affect the organizational effectiveness of availability management. So may a change of the staff level. Both can affect the expectations of availability performance. The result will be reflected by reduced maintainability in the availability formula (Equation 1-1).

Previously identified tasks and responsibilities may no longer be appropriately assigned. There may also be inadequate numbers of personnel to do the work. This may directly or indirectly undermine availability functions and tasks.

Management has the choice of accepting the consequences or adjusting the availability scheme. The choices are as follows:

- Accept the consequences if the net change in productive capacity and its economics are still positive.
- Redesign plant reliability and maintainability to achieve acceptable availability performance. This is especially possible through the plant maintenance operations.
- Reconsider the initial change if neither situation applies.

Baseline Design

Baseline design and its control is fundamental to maintaining the integrity of the availability scheme. Therefore, baseline design is shown in Figure 5-3 as central to improvement, change, and data management. This is because improvement and change depend on the knowledge of what currently exists (a baseline), holistic evaluation of the proposed changes against the baseline, and controlled execution of the change.

Data Management

Availability data management is the foundation or enabling capability for improvement, change, and baseline management.

Vision of the data management system. The vision of an availability data management system is simple. All models, management tools, databases, and documents associated with the plant's designed and managed availability scheme are captured and integrated within the overall plant computer systems. All or most of these are placed in electronic form. Thus, the data management system is the residence of the baseline availability scheme as a living design to be revised as plant performance needs and conditions change with time.

The purpose of data management. The data management system is an important part of availability engineering and management. This is because plant management is responsible for the acquisition, evaluation, improvement, and revision, operation, and support of the plant. Such responsibility requires feedback.

Feedback is necessary to establish the reliability, maintainability, and economic parameters of availability management. It is also necessary to monitor and control maintenance operations performance.

To this end, large amounts of raw data must be generated. However, it must be reduced and refined to an appropriate quantity and form to be useful to management.

Thus, data management is the collection, storage, retrieval, and processing of data into the living availability design. It allows management to regularly manipulate data to generate reports on all types of availability performance. Its operations research models can be drawn upon to produce special reports and to solve problems. In fact, the only limit on the analytical processes and reports is what has been and will be collected.

New systems and plant system integration.   The data management system is not necessarily a grand package of hardware and software to be installed. It is a mixture of programming, off-the-shelf packages, and systems integration. Accordingly, it adds components to existing systems associated with engineering and management availability design. These are integrated to complement existing systems.[1] They both draw upon and provide input to common databases.

Determining availability data requirements. Data requirements for availability management must be determined by an orderly process. How availability and maintenance operation performance will be measured was established by the conceptual phase availability deliverables. The steps to determine their respective data needs are as follows:

- Identify the information products required to evaluate the identified performance measures.
- Determine the necessary calculations and models to generate the information products.
- Identify the associated data inputs.
- Evaluate the sufficiency of data collected by existing or proposed data systems.

Data to account for maintenance time and requirements.   The previous section described the steps for determining data requirements. The top-level data and information requirements associated with maintainability are:

---

[1] Existing systems commonly include material inventories and stock cataloging; preventive maintenance and equipment records; work-order costing; equipment inspection; planning and scheduling major maintenance projects; and work-order planning and scheduling.

- A method for identifying the equipment under study.
- Why the equipment required maintenance.
- When the maintenance requirement became known.
- How the maintenance requirement was detected.
- A method of linking a series of related maintenance and support tasks and resources stemming from the original maintenance requirement.
- The amount of time required to undertake the subject maintenance activity.
- What was done to restore the equipment to an operating state.

Data to account for time between failures and consequences. The above will account for the maintenance time and requirements. There must also be an accounting of the time elapsed between failure and its consequences. The following fulfill that need:

- A method of identifying equipment operating time.
- A method of linking equipment to its operating mode.
- An ability to identify an operational discrepancy with narrative information when necessary.
- An ability to identify when the discrepancy occurred.

Timing to Implement the Functions and Systems

The improvement, change, and data management functions and systems must be assessed and planned as part of the design phases. They must also be implemented during these phases. The objective is to have them in place by the end of plant startup. The interim objective is the timely capture of the evolving baseline design.

It is easy to visualize baseline design as a completed set of calculations, design, and construction documents. However, the project milestones will produce design results that must be protected as a baseline. Therefore, the permanent improvement, change, and data management functions and systems will begin to evolve during the design phases.

This may go unnoticed as part of the traditional project document control practices. Historically, the project function has been regarded narrowly. Much of its final content is expected to ultimately be sent to a static archive instead of becoming a living design.

Tasks to Accomplish the Deliverable for Improvement, Change, and Data Management Functions and Systems

The tasks to develop the improvement, change, and data management functions and systems are described in following sections. They are flow-charted in Figure 5-4. The flowchart includes the sequences and iterations with tasks in parallel and previous deliverables.

Task legend (basic design phase):
AB1.3 Define role of functions in availability factors.
AB1.7 Make practices development decisions.
AB3.7 Define system for continual collection & processing of data.
AB4.5 Define system to manage plant logic diagrams & allocated availabilty parameters.
AB6.5 Develop plans for availability model.
AB7.7 Develop plans for financial model.
AB8.2 Identify all failures for FMECA.
AD1.2 Plan & organize for acquistion & development of maintenance task analysis detail.
AD2.1 Define functions, library & system for technical materials.

AD2.3 Define organization for techinical materials management.
AD4.2 Define & develop system for resource modeling.
AD5.2 Define & develop system for facility modeling.
AD6.10 Define organization for training program.
AD7.1 Process & linkage diagrams for maintenance operations.
AD7.4 Systems selection & integration for maintenance operations.
AD8.2 Process & linkages of availability management.
AD8.7 Define and detail the overall availability management organization.
AD8.8 Define systems to retain detail of availability organization.

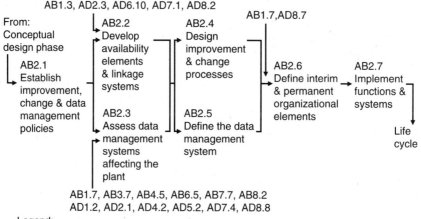

Legend:
AB = Availability deliverable or task in the basic design phase.
AD = Availability deliverable or task in the detailed design phase.

Fig. 5–4 Tasks to plan and organize the availability-centered improvement, change and data management functions and systems (AB2 of Figure 5-1).

AB2.1. Establish improvement, change, and data management policy. The management of improvement, change, and data requires direction and careful planning. This is because considerable human, hardware, software, and, therefore, financial resources are required. Furthermore, the results are extremely important to the owner's business success.

Therefore, management should establish policies for the following:

- The subject types of changes and improvements.
- How these changes and improvements will be processed.
- The responsible organizational entities to be involved in that process.
- How the success of these policies will be measured.

The policies should be correlated with other corporate and plant initiatives. For example, there may be a program to institute computer integrated manufacturing (CIM). The availability system should be an integral part of such initiatives. This is because availability engineering and management is a major dimension to be served by computer integrated manufacturing.

AB2.2. Develop availability element and linkage systems. It is important to carefully map the elements of availability and their linkages to each other. The purpose is to develop the ability to trace and cross reference functions, controls, procedures, and existing design. Therefore, a system to define elements and their linkages is an important tool.

The elements of availability management should be carefully identified. The inputs from suppliers along with outputs to customers are the linkages of each. Owners must also be identified.

The deliverable for developing availability-centered practices control documents (AB1) contributes to this identification. This is shown in Figure 5-4. The organizational functions relevant to availability performance were determined (AB1.3). The findings provide information to help determine the suppliers, customers, and owners.

An ideal situation would be to form diagrams that can be followed like a road map to thoroughly trace the consequences of plant changes to identify availability engineering and management elements that require realignment.

The principle is indisputable. However, lucidity may not be possible because of the inherent complexity of a typical plant. The diagrams may be either too simple or too much of a maze to be useful.

The problem can be circumvented with program management software that networks all of the program's elements. It can be designed to cross-reference or link the functions, controls, procedures, practice documents, and designs affected by a change. It can also access and status these elements from all directions. Figure 5-4 shows that the availability element and linkage system will become more detailed with time. The detail of other organization subsystems will be defined by other deliverables. The tasks that provide this detail are as follows:

- Definition of organizational processes for technical materials management (AD2.3, Chapter 11).
- Definition of organizational processes for training program management (AD6.10, Chapter 12).
- Development of processes and linkages for maintenance operations (AD7.1, Chapter 13).
- Development of processes and linkages of overall availability management (AD8.2, Chapter 13). This task refines the first three tasks.

AB2.3.   Assess   the   data   management   systems   surrounding   the plant.   Any plant is surrounded locally and organizationally by existing and planned data management systems. Accordingly, this task is to identify and assess those that could influence the availability scheme. The assessment should include the following:

- The strengths and weaknesses of each data management system.
- The capacity of existing systems to be interfaced with new database schemes.
- Management's explicit and implicit intentions to develop future systems.
- Management practices served by these data management systems.
- What data and information is normally collected, its form, how it is processed, and its current and future uses.
- Indirect possibilities that could be served by the systems. For example, how can they help transfer accumulating availability knowledge, experience, and methods throughout the owner's enterprise?

AB2.4. Design the availability improvement and change processes.   The management of change goes beyond the need to identify a desired improvement or change, acquire funds, and make the change. Therefore, its processes should be carefully developed. Some basic development issues are:

- Procedures, practices, and controls for identifying and managing changes.
- Information and documents that will be used to support the decision-making processes.
- How to collect the information that describes what decisions were made and why.

- Document controls associated with the processes of the improvement, change, and data management systems.
- How the data management systems will support the efficient creation, revision, and management of maintenance and administrative procedures.
- Programs to train employees in the concepts, purposes, goals, and practices of the improvement, change, and data management functions and systems.
- How the trails leading to the current status of plant availability elements will be developed for future review and referencing.
- Mechanisms for obtaining, retaining, evaluating and distributing feedback from field and support personnel.
- Procedures to trigger and guide the continuous assessment of accumulating plant availability experience.

AB2.5. Define the data management system.   Defining the data management should include the following:

- A configuration diagram of hardware, software, and databases as a system.
- Data and information products generated from the system components.
- An interface scheme to integrate the subsystems. The subsystems and their sources are identified as input tasks.
- A list of users of the information product.
- Organizational functions as the owners of each system component and product.
- Selection and design criteria for each system component.

Earlier in this chapter it was mentioned that the data management system is the residence of the living availability design. Thus, various databases and models are captured and integrated as a system. In Figure 5-4 it is shown that the following tasks of other deliverables flow to this one:

- Availability-centered practices control documents are text data to be assessed and managed. They are the outcome of the decision for which practices are to be developed (AB1.7, Chapter 5).
- A system to gather and analyze data throughout the plant's lifetime will be defined (AB3.7, Chapter 6). Its long-term product is a continually improving database of reliability, maintainability, and economic data.
- Plant logic diagrams are developed in another deliverable (AB4, Chapter 6). The logic diagrams and developed data are ultimately synthesized as a system of allocated reliability, maintainability, and eco-

nomic parameters. When the logic diagrams are first developed, a system must be defined to retain and access their synthesis (AB5.5).

- Availability and financial models will be developed for plant design and management (AB6 and AB7, Chapter 7). Each will prepare plans for developing its associated model (AB6.5 and AB7.7 respectively).
- FMECA and the subsequent maintenance logic-tree analysis (AB8 and AB9, Chapter 8) will search out potential significant failures and determine appropriate maintenance strategies. One task is to define the integrated data system to access, use, and revise these analyses throughout the plant's life (AB8.2).
- The data systems will be defined to serve the availability models, databases, and associated systems that will take form in the detailed design phase. They are the subject of Part 4. They include systems for determining the following:

    Maintenance task analysis (AD1.2, Chapter 11).

    Maintenance procedures (AD2.1, Chapter 11).

    Calculation of maintenance resources (AD4.2, Chapter 12).

    Systems used in maintenance operation functioning (AD7.4, Chapter 12).

    Determination of support facilities (AD5.2, Chapter 12).

    Systems for training program design and management (AD6.11, Chapter 12).

    Systems to manage the detail and tools of the availability management organization (AD8.8, Chapter 13).

This list suggests a problem of analysis. The system incorporates subsystems developed in later deliverables. They are developed to manage major functions and elements of availability performance.

This problem is solved in the conceptual design phase. A task in the availability concept will generally define the full set of computer systems and their integration (AC2.10, Chapter 3). Consequently, the later inputs to this task will be refinements rather than initial determination of subsystems.

AB2.6. Define interim and permanent structure, authority and responsibility. A definitive structure of authority and responsibility must be established with respect to planning, implementation, and ultimate management.

The structure should be metamorphic. The need for improvement, change, and data management begins with the earliest design phases and continues until the plant is retired from service. As it passes through life, its organizational needs change. In turn, the change is forced upon the structure of authorities and responsibilities.

The structure should recognize two basic stages:

- Planning and implementation during the plant design, con truction, and startup phases.
- Management as a function in the plant's producing life.

With respect to data management, the first manages the process of system definitions, physical design, and implementation. The second manages its use, enforces its procedures, maintains its integrity, and advances its abilities.

The structure is derived rather than formulated by managerial intuition. The necessary process of organizational design is described in Chapter 13 (AD6). Therefore, the process will not be addressed here.

In Figure 5-4 it is shown that the organization scheme has two inputs. They are as follows:

- The decision to develop a system of "proactive" control documents is managed by the improvement, change, and data management organization. Thus, the final development decision (AB1.7) is an input to this task.
- The final design of the overall availability organization (AD8.7, Chapter 13) includes this organization. Thus, the final organization design will be refining input to this task.

AD2.7. Implement the improvement, change, and data management functions and systems.   The previous deliverables have defined what is necessary for improvement, change, and data management. They have also defined the organizational entities to be responsible. Thus, this task, as a general one, is to accomplish that scheme.

This deliverable is a substantial and specialized program in its own right. Therefore, no attempt will be made to describe it here.

Bibliography

Blanchard, Benjamin S. *Logistics Engineering and Management*. 2nd ed. Englewood Cliffs, N.J., Prentice Hall, 1991.

Center for Chemical Process Safety. Guidelines for Technical Management of Chemical Process Safety. American Institute of Chemical Engineers. New York, 1989.

Davis, Gordon B. Management Information Systems: *Conceptual Foundations, Structure and Development*. 2nd ed. McGraw Hill, New York, 1985.

Du Bois, D'Andrea; Ostrofsky, Benjamin; and Arnold, T. S. *Systems Effectiveness Data System.* Volume I; Management, Analysis and Programs. Redondo Beach, Calif. TRW Systems Group, August, 1969.

Pack, R.W., Seminara, J.L., Shewbridge, E.G. and Gonzalez, W.R. Human Engineering Design Guidelines for Maintainability. Atlanta, Ga., General Physics Corporation, 1985. EPRI NP-4350.

Sanders, Mark S. and McCormick, Ernest J. *Human Factors in Engineering and Design.* 7th ed. New York, McGraw Hill, 1993.

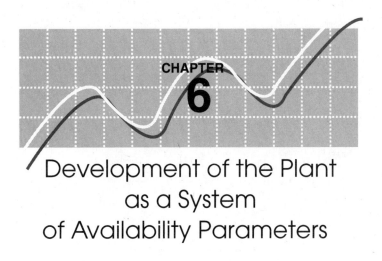

Development of the Plant
as a System
of Availability Parameters

Introduction

The previous chapter was concerned with the infrastructure to maintain and advance the integrity of plant availability performance. This chapter describes the next basic requirement: the development of the plant as a system of reliability, maintainability, and economic performance parameters.

Traditional design includes production-process flow diagrams for plant design in which performance parameters are specified for each point in the production process. There is an equivalent requirement for plant availability performance. However, the associated plant diagrams specify the parameters of reliability and maintainability. Both cases include associated economics.

The deliverables for developing these availability parameters and diagrams are bounded in Figure 6-1. The figure includes the three deliverables that are the subject of this chapter. They are as follows:

- Identifying, gathering, and analyzing data for availability design (AB3).
- Developing plant logic diagrams (AB4).
- Allocating availability parameters to the plant logic diagrams (AB5).

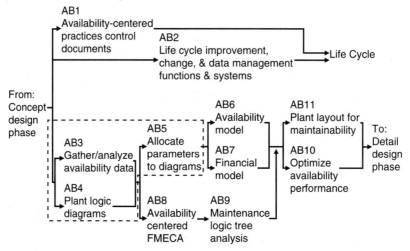

AB = Availability deliverable in Basic design phase
FMECA = Failure modes, effects & criticality analysis

Fig. 6–1 Basic design phase availability engineering deliverables to develop the plant as a system of availability parameters.

AB3. Identify, Gather, and Analyze Data for Availability Design

Definition and Purpose

The determination of availability performance requires data. The purpose of this deliverable is to gather and develop the raw data to be analyzed and reformulated for application to the subject plant.

The Goals, Philosophy, and Nature of Data

The development of data is a challenge. One reason is that availability performance has not been designed and managed in the past. Although the data may be plentiful, there may be little history of managing and reforming it to an actual end.

Therefore, it is important to consider the philosophical nature of the problem. Otherwise, data may be regarded as an unsurmountable obstacle rather than just one challenge in a larger quest. Management can ill afford to forsake possibly hundreds of millions of dollars in possible lifetime income because all the necessary data may not be perfect.

Return on investment in data. Many owners have incurred great cost over many years to capture plant data. This includes the cost of hardware, software, and procedures to gather it. Furthermore, the cost of its management is a continual operating expense against income.

Engineering and management of availability performance provides an opportunity to capitalize on this considerable investment. A return on its investment comes from its application to initial plant design. This return continues throughout the plant's life as it:

- is improved by focused collection and analyses over the plant's lifetime,
- is used to continuously improve management's ability to predict and optimize availability performance,
- becomes part of the living availability design applied in response to changing trends in the demand for productive capacity,
- is available to future capital projects in the plant,
- is made useful to traditional corporate, business, and plant-level cycles of general business and production management, and
- is made available to availability design, improvement, and management in other plants and capital projects.

Definition of adequate data. It is important to define what is adequate data for initial availability design. Otherwise, the design goal can be defeated by an unrealistic quest for perfect rather than effective data.

Performance levels and their analyses in the plant have two extremes. At one extreme is equipment, subassemblies, and components. At the other is plant availability performance. This is an important distinction because the definition of appropriate design data is different for each.

Data must be gathered and developed for each plant element. However, design for the performance of equipment and lower-level elements benefit the most from rigorous collection of statistical data. The design of plant-level performance does not depend on such perfection. Instead, it requires realistic forecasts of performance for each significant plant element. They are produced by combining expertise and experience with available data.

One reason for the distinction is statistical reality. Plant- level performance is the synthesis of the reliability, maintainability, and economic forecasts of many elements. The system-level prediction gravitates to some central result and confidence limits. Consequently, immense time and cost to initially acquire perfect data is not a prerequisite to successful initial availability design. Thus, the initial stress on perfection is focused on the most critical items.

This is not to suggest that rigorous historical data is not valuable. In the short-term, it is a means to confirm and refine the forecasted performance of plant items. In the long-term, it is an ideal to which to aspire.

Our paradigm of data is, perhaps, a part of that challenge. Many engineers and other experts have spent a career designing and solving the problems of just a single or several pieces of equipment. Some have even specialized in components and subassemblies. When faced with system-level analysis and design, it is easy not to concurrently shift our paradigm of what constitutes appropriate data.

Data for the short- and long-term.   The rigorousness of available data should not be viewed as a permanent condition because availability performance is not just a problem to be solved during the initial design phase. Instead, management must be concerned with performance as a living solution. Plans should be made for the data to be progressively improved as part of the larger goal. Even then the goal is not data. It is to continually improve management's ability to predict, improve, and better optimize plant availability performance.

Chapter 2 introduced the concept of operating leverage. That is, a percentage of change of availability or its cost structure will produce a significantly greater change in income. This leverage is greatest for plants that require a high-capacity utilization rate to generate income.

Perfection in the practice of availability design is not critical to creating immense consequences for income. Operating leverage suggests that immense financial ramifications result from doing the right things. The perfection of these practices comes with time. This is so only because perfection was defined and made possible by doing the right things. Furthermore, it is possible that the incremental income from perfection will never match the initial payoff of doing the right things.

This can be translated to the challenge of data. "Doing the right thing" is finding approaches to reasonably predict reliability, maintainability, and economic performance of plant items. Perfection is developing rigorous historical data. What, when, and how it is developed over time is defined by the overall approach to availability engineering and management. Meanwhile, there will be immense financial benefits from doing what is necessary to develop an effective initial availability design.

Sources of Data

The primary data sources for availability engineering and management are:

- Similar plants, subsystems, or equipment assemblies and components in similar operating environments.
- Manufacturer's design experience, tests, and evaluations.
- Published data.
- Consensus and analysis by people with experience and expertise in the operation, maintenance, and analysis of similar equipment.

Each of these is discussed in the following sections. Figure 6-2 suggests that they are drawn upon in an integrated approach to data development. Consensus and analysis is shown in the center. This is because of its role in making the other sources possible. It is the dimension of "real" intelligence in the development of data.

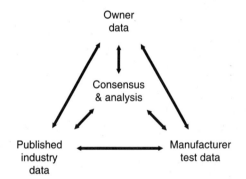

Fig. 6–2 Sources and integrated relationship of data and information for availability design and management.

Data from Similar Plants, Subsystems, and Equipment

Existing facilities offer three main sources of data. They are:

- Plant production process control systems that routinely collect massive amounts of data.
- Maintenance operations that produce data associated with the time and resources to repair equipment.
- Predictive maintenance procedures and systems that collect data in the process of monitoring the condition of plant equipment.

The data is sometimes available in electronic form. Thus, the electronic extraction and manipulation of large amounts of data is possible.

Manufacturers as a Source of Data

Another possible source of information is manufacturers. However, this source presents its own special challenges. The people who are most accessible are not always technically knowledgeable about the data being sought. They may not even realize that their organization has the desired information. Nor will they always understand the nature of the intended application.

Experience has shown that highly qualified personnel must be assigned to gather data from manufacturers. One reason is their ability to lead the contact to their own data. Another is their ability to determine what the offered data actually represents.

Published Industry Data

The third possible source is published industry data. This is the least attractive source. Therefore, it should be the last resort.

Weaknesses of industry data.   The weaknesses of this source are as follows:

- It is the aggregate of different and generally unknown operating conditions.
- Because of this inconsistency, there is no reasonable way to realistically adjust the data to planned operating conditions.
- The capacity to form reasonable failure and maintenance distribution functions is accordingly limited.

These limitations are dramatized by the experience of those who have used this data. They have found huge differences in the various industry databases.

When the source is relevant.   The pertinent question is: Why use this source? Possible reasons are:

- Preliminary assessments and comparisons. These may test a design, help determine relative importance of elements to plant level performance, etc.
- Plant elements that preliminary availability modeling has found to be insignificant.

Consensus and Analysis by Experience and Expertise

Data derived from the consensus and analysis of people with experience and analytical expertise is very valuable. Such immense talent often exists within the owner's organization. It is also available from businesses that provide specialized consultants and from some project contractors.

Teams specializing in various types of equipment are usually necessary to integrate experience and analytical abilities. One member of the team should be an engineer who is an expert in the necessary analytical techniques. Another should be an experienced "hand" who has seen the consequences of different operating conditions and practices in the field.

The general expectation is that these teams can formulate reasonable forecasts of time-to-failure, time-to-maintain, and their associated economics. The teams should also be able to establish a sense of confidence limits.

Teams are important sources of data because of their nature as "real" intelligence. Consequently, they provide a capability that the other sources cannot. The teams can make predictions for different operating conditions. The other sources can only reflect historical conditions. Therefore, they are regarded as raw data. Teams are regarded as the means to convert the raw data to the subject facility.

Teams are attractive sources for another, less direct reason. They formally draw personnel from nontraditional disciplines into the plant design. Examples are maintenance, reliability, and operating personnel that typically have participated only later in the plant's life. By that time, their tasks were to rectify problems left or even created by the initial design. Participation in the design phase tends to be limited and unstructured.

An Integrated Approach to Data and Information

The data sources and their treatment are approached as integrated rather than independent possibilities. This is depicted in Figure 6-2. The driving issue is the relative value of specific data items to the design and management of availability performance.

Preliminary availability modeling helps guide the data scheme and reveals the relative importance of data voids and uncertainties in the availability calculation. The design team can then determine the importance of finding solutions. Preliminary modeling also reveals where not to expend energy beyond the basic processing of readily available data.

Consequently, at one extreme data development may be limited to published data. Other data requirements may be collected from expert and manufacturer sources and then subjected to rigorous analyses. A further extreme is data derived from testing programs. All of these approaches may be applied progressively over time to the same requirements.

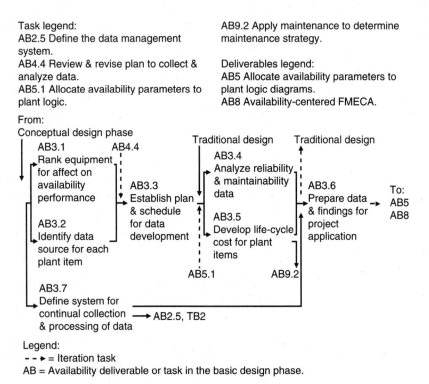

Fig. 6–3 Tasks for planning and developing availability design (AB3 of Figure 6-1)

Tasks for Accomplishing the Deliverable to Plan and Develop Availability Data

The tasks of this deliverable are flowcharted in Figure 6-3. The flowchart also shows their sequences and interactions with tasks in other deliverables.

AB3.1. Rank equipment for their affect on availability performance. The first task is to rank the importance of equipment, assemblies, and components with respect to reliability, maintainability, and economics. A preliminary availability model was formed during the conceptual phase. Its results are used here as an assessment tool. The objective is to establish the relative importance of these elements rather than to produce an exacting or sophisticated ranking result.

AB3.2. Identify the data sources for each plant item. The specific sources of reliability, maintainability, and economic data must be identified for each plant item. The task is also to determine any anticipated difficulties in collection and analysis.

AB3.3. Establish a project plan and schedule for data development. Project planning for the development of data is important because of the magnitude of the work and the need for timely results.

The planning requirements are as follows:

- Establish categories of analysis with respect to rigor.
- Develop a basic analysis approach for each category.
- Determine the schedule for the collection and transformation of data for project use. The schedule is derived backward from key availability and traditional project milestones.

Figure 6-3 shows that this task may iterate with the deliverable for developing plant logic diagrams (AB4). This will be with the task to review and revise the plans for collecting and analyzing data (AB4.4). The team that developed the plant logic diagrams may both add and eliminate elements. The team may also revise the desired quality of specific data elements.

AB3.4. Analyze reliability and maintainability data. Reliability and maintenance specialists should carry out the planned analyses. Others, such as mechanical engineers, chemical engineers, metallurgists, etc., should also be involved. The product is forecasted reliability and maintainability performance. Conclusions will be in terms of means and range of occurrence.

Some issues to be explored by the interdisciplinary team while forecasting performance are:

- Identifying when improved components, materials, etc., would affect reliability and maintainability, and their economics.
- Identifying weaknesses in the item.
- Discovering important uncertainties during the item's life and determining whether they could be controlled.
- Determining the possibilities and consequences of simplification in terms of number of parts, precision, necessary maintenance skills, etc.
- Determining how the item's service life would be reduced by abnormal operating conditions.[1]
- Identifying the design and construction possibilities that may affect reliability. Examples are poor installation, foundations that allow vibration,

---

[1] Such conditions are forces such as mechanical loads, deflections, and pressure; reactive agents; environment; temperature; and the time of exposure to these conditions.

incorrect equipment sizing, poor casing support, piping strain and support, etc.

- Determining what the previous findings suggest are the potential consequences of departing from the traditional experience. Is less plant needed than historical common knowledge thought necessary? Are opportunities revealed to exceed the performance of predecessor plants?[2]

AB3.5. Develop life-cycle cost data. The previous task developed reliability and maintainability data. This task is to develop the life-cycle cost data associated with those findings. It is also to develop a cost breakdown structure according to life cycle phases. The cost categories must be rigorously determined for each phase. Blanchard and Fabrycky [1990] provide excellent detailed lists of life-cycle cost items which are used to form a cost list for each phase. The resulting cost breakdown structure is ultimately applied to each financially significant item.

Figure 6-3 shows that this and the previous task take general input and then iterate with the traditional design phase deliverables. All traditional design deliverables are conceivably interfaced with availability engineering. The most obvious relationship is between the final conceptual design (TB1) and the performance specification, sizing, and selection of equipment (TB4). The design decisions made during this phase affect the operating conditions created in the plant and determine the plant's maximum capacity for availability performance.

The task is also shown to provide direct input for the selection of maintenance strategies (A9.2, Chapter 8). This is because some strategies are selected based on economics.

Figure 6-3 shows the iteration from a later task to allocate reliability, maintainability, and economic parameters (AB5.1). This is the case for this and the previous task (AB3.4). The revised parameters are, in turn, the result of other deliverables in the detailed design phase.

The specific source of the iteration is the maintenance task analysis. Its tasks include networking and detailing the active repair, logistic, and administrative elements of each maintenance task.

This task and the preceding one will initially estimate these elements and the time to return an item to service. The maintenance task analysis will no doubt refine them. As the refinements are allocated to plant logic, they flow to this and the preceding task.

---

[2] This suggests that each future capital project or initiative for performance advancement will capitalize on the quality of availability engineering and management practices that have gone before them.

AB3.6. Prepare data and findings for project application.   The findings and data are organized and documented. The objective is to prepare them for use in the availability engineering process. It is also to make them available to cycles of access and revision throughout the plant's life. Accordingly, the work product must be entered in the data management system described in the next section.

AB3.7. Define a system for the continual collection and processing of data.   The preceding task may have created a sense that data development is a design-phase activity. However, data development continues throughout the plant's lifetime. The goal is to continually improve management's ability to forecast plant performance.

This task is to define the system for that purpose. The issues are as follows:

- How is availability performance to be measured, managed, and improved?
- What data products are required?
- What are the analysis methods and formulas required to generate them? These may be defined for the plant's life cycle. As data accumulates, the planners may wish to have applied increasingly sophisticated methods. These decisions are especially important to the strategy for the long-term development of current weakly formed or unavailable availability performance parameters. These strategies will be formulated in a later task (AB5.3, Chapter 6).
- What are the required data elements and their sources?
- How are they to be continually collected or accessed by the data development system?
- What are the access, collection, and analysis cycles?
- What human analysis of the data gathered or accessed is to take place?

The reader will notice that some of the issues are at least partially answered by the preceding tasks. Other deliverables also provide source information. The availability and maintenance operation concepts (AC2 and AC3, Chapter 3) identify the basic measurements of performance.

Many of the other availability deliverables define database and model requirements to be integrated in the overall availability-centered data management system. The results of this task also flow to that definition (AB2.5, Chapter 5). Thus, they will be integrated with them.

The results of this task also affect the design of the plant process control system (TB3, Chapter 10). They identify the data to be acquired as a normal part of process control (SCADA).

Probability Distribution Functions for Data

The primary goal of this deliverable is to ultimately convert data to reliability, maintainability, and economic performance parameters. They are expressed as probability distribution functions rather than cardinal expressions such as mean time-to-failure, etc. Therefore, the analyst's final role is to formulate data into appropriate distribution functions.

Variable failure patterns. There are various probability distribution functions. They reflect patterns of failures.

For example, the distribution functions for electronic parts is substantially different than those for mechanical parts. Mechanical failure rates increase over time while electrical and electronic components failure rates remain constant. Meanwhile, mechanical items have multiple failure patterns. The characteristics of failure will vary with time as well. They are also much more sensitive to operating conditions.

Types of probability distributions and their application. Possible applications of probability distribution functions are as follows:

- The exponential distribution is applicable to electrical and electronic elements. They are subject to random failure events. Thus, they experience constant failure rates.
- The normal distribution is applicable to three cases. First, for failures from thermal creep and wear. Second, when a large number of elements and their failures are combined instead of being treated individually. And third, for time-to-maintain in relatively simple tasks that experience little variation in their scope and time requirement.
- The lognormal distribution is applicable to failures from cracks and corrosion. It is also applicable to complex repair requirements that experience relatively variable time-to-maintain.
- The Weibull distribution may be used to analyze and forecast mechanical failures. It reflects three stages of life: run-in or infancy, normal useful life, and final wearout. The first and last period have higher failure rates. The middle period has random failures.

The preceding functions are not presented here as a prescription. In fact, discussions with practitioners in the field of probability and reliability analysis show that the choices are variable and complex. The message of this discovery is that the availability design team should recognize the importance of these functions and include these practitioners in the design team.

AB4. Develop Plant Logic Diagrams

Definition and Purpose

The logic diagrams depict the plant elements as a system. Reliability, maintainability, and economic parameters will ultimately be assigned to each of the plant's equipment, assemblies and components. This becomes the foundation upon which later availability and financial models are built.

Comprehensive Function Logic Diagrams

The inclusive approach.   The United States Department of Defense approaches this deliverable inclusively. Both the hard plant and its associated activities are diagramed. The latter spans design, construction, testing, deployment, operating cycle, field maintenance, support, and training.

The prescribed method is to diagram all plant aspects. All that happens in the subject system is called functions. The prescribed approach is to start at the top level and progress downward. Only at the final step, are equipment items identified for each pertinent function in the production process.

Equipment as the logic diagram.   The inclusive approach is not particularly practical for production plant design. The equipment that serves process functions is typically identified almost immediately. In other words, if there are options for grouping production steps, they are limited. The immediate issue is whether the configuration and choices of equipment types can be expected to achieve the specified plant availability performance. Furthermore, decisions to package sets of process functions as an equipment item are generally predetermined.

When plant function logic is formulated.   The principle of plant activities analysis and diagramming is not omitted. Instead, it is fulfilled at other times during availability design. These cases are as follows:

- Pertinent plant management functions, their linkages, procedures, etc. are diagrammed to develop the availability improvement, change, and data management functions and systems during the basic design phase.
- Maintenance procedures are developed during the detailed design phase. These are preceded by the determination of significant failures and appropriate maintenance responses. Each maintenance task is detailed as a network of steps for active repair, administration, and logistic support.

- Availability, maintenance operation, and training functions and organizations are designed in the detailed design phase. The objective is to develop an effective overall management organization.

Plant logic diagrams versus function logic diagrams.   Because of their more direct approach, the diagrams in this deliverable are called *plant logic diagrams* rather than *function logic diagrams*. They are formulated from the various traditional engineering diagrams and details. Thus, the logic is generally limited to hard plant elements.

Types of Plant Logic Diagrams

Two types of diagrams.   There are two types of plant logic diagrams. The simplest is the block flow diagram. The most complex is the Boolean algebra logic diagram. It depicts the plant in the logic of AND, OR, NAND, NOR, and NOT.

Advantages and disadvantages.   Both types have advantages and disadvantages. Block diagrams make it easier to understand the network of system components.   This is demonstrated in Figure 6-4. The same simple system is shown in block and Boolean form.

The simplicity of block diagrams is an advantage when an interdisciplinary team first develops the plant logic. It is even more of an advantage when nontraditional personnel are involved. Its simplicity allows the team to concentrate on understanding the plant rather than the method.

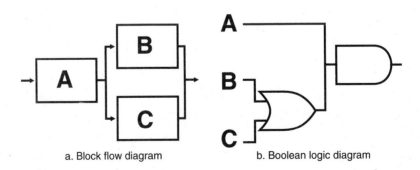

a. Block flow diagram   b. Boolean logic diagram

Fig. 6–4 Diagraming methods; function flow and boolean logic approaches.

Boolean logic provides much more information concerning the role of elements in the plant as a system. For example, Figure 6-4 shows items "A" and "B" as two parallel components. However, do they function at the same time? Or do they function as redundant items? The Boolean logic of Figure 6-4.b depicts the operating relationship. The function block diagram cannot. Therefore, Boolean logic can better analyze availability performance with various operating and failure scenarios.

When to use them. It is probably most effective to use both types of diagrams. The timing and nature of their use are as follows:

- Block flow diagrams are used to initially formulate the plant diagram and allocate parameters to it. When these are drawn, many people will be involved. Consequently, the issue is the group's capacity to communicate, visualize, and grasp plant relationships.
- The completed block diagrams are converted to the Boolean format to serve modeling needs. Typically, few people are involved in this stage. Thus, complexity is not a problem. The issue is analytical power.

Tasks to Accomplish the Deliverable to Develop Plant Logic Diagrams

The deliverable to develop the plant logic diagrams requires the tasks described in the following sections. They are flowcharted in Figure 6-5. The flowchart also shows their sequence and iteration with tasks in other deliverables.

AB4.1. Gather pertinent existing design products. This task is to gather all pertinent plant design products. These include the following as they become available:

- Various traditional engineering diagrams which have been formulated to depict the plant process, configuration, and equipment. This input is shown in Figure 6-5.
- Engineering documents that describe how the plant is to work.
- Availability engineering documents that identified plant subsystems with respect to their relevance to plant availability performance. These were formulated in the conceptual design phase.

AB4.2. Establish the set of plant diagrams. This planning task is to determine the complete set of necessary plant logic diagrams. These are the previously mentioned block flow and Boolean logic diagrams. From these,

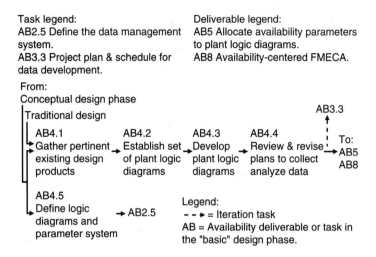

Task legend:
AB2.5 Define the data management system.
AB3.3 Project plan & schedule for data development.

Deliverable legend:
AB5 Allocate availability parameters to plant logic diagrams.
AB8 Availability-centered FMECA.

From:
Conceptual design phase

Traditional design

AB4.1 Gather pertinent existing design products

AB4.2 Establish set of plant logic diagrams

AB4.3 Develop plant logic diagrams

AB4.4 Review & revise plans to collect analyze data

AB3.3

To: AB5 AB8

AB4.5 Define logic diagrams and parameter system

→ AB2.5

Legend:
- - ► = Iteration task
AB = Availability deliverable or task in the "basic" design phase.

Fig. 6–5 Tasks to develop plant logic diagrams (AB4 of Figure 6-1).

planning and scheduling for this and subsequent deliverables are accomplished.

AB4.3. Develop plant logic diagrams. This task is to develop the planned logic diagrams. As representatives of technical and management disciplines, the participants must determine the appropriate level of resolution that was initially established in the conceptual design phase. It may be revised during this task to reflect a growing knowledge of plant design and desired performance.

Equipment does not fail, its parts do. Therefore, the plant logic diagram should go beyond just the configuration of equipment. Equipment should be detailed as a network of primary components or assemblies. The resolution should be a function of elements that experience has found to be probable sources of significant maintenance requirements.

AB4.4. Review and revise the plan to collect and analyze data. The previous task may involve adding or deleting elements. It may also change the level of resolution or desired confidence limits of data. Thus, reliability, maintainability, and economic parameters must be formulated for each element in the diagram. This is the purpose of the parallel deliverable of developing availability data (AB3, Chapter 6). Therefore, the task may iterate with the task to plan and schedule the development of data (AB3.3).

AB4.5. Define the system to manage the plant logic diagrams and allo-cated availability parameters.   The plant logic diagrams and their parameters (AB5.5) will ultimately be a part of the living availability design. Thus, this task is to define a system to manage, access, and modify the results. This is necessary for the initial plant design and for the subsequent improvement and change over its life.

Figure 6-5 shows that the result of this task flows to the deliverable to develop the availability-centered life-cycle improvement, change, and data management functions and systems (AB2, Chapter 5). It is incorporated in the defined overall data management system (AB2.5)AB5.

AB5. Allocate Availability Parameters to Plant Logic Diagrams

The purpose of this deliverable is the synthesis of the previous two deliverables. The result will be logic diagrams with performance parameters assigned to each element. The assigned parameters will later be input to the availability and financial models.

Compensating for Unavailable and Weak Data

It was suggested earlier in this chapter that data development may be a challenge. Thus, allocation of reliability parameters may need to compensate for unavailable and weak data. This may occur when the plant incorporates new process and equipment technology. Another reason may be that the data has not been efficiently collected in the past. Although data may be important, the time and cost of its acquisition may be prohibitive. These constraints can be overcome as follows:

- Allocate known detail to the plant logic.
- Compute the availability significance of the plant element for which data is unavailable. In other words, what do they represent in the over-all calculation of availability?
- Make an allowance to availability performance for the unknown reliability and maintainability parameters. This will be a function of criticality, number, and relationship of the associated parts of the elements and their surrounding operating conditions.
- Formulate plans for how these allowances will be validated and refined as data accumulates during the plant's design phases and producing life.

Tasks for Accomplishing the Deliverable to Allocate Availability Parameters to Plant Logic Diagrams

The tasks for allocating reliability, maintainability, and economic parameters to the plant logic diagrams are described in the following sections. They are flowcharted in Figure 6-6. Also shown are the sequences and iterations with tasks in parallel and previous deliverables.

AB5.1. Allocate reliability, maintainability, and economic parameters. Parameters are assigned to the plant logic diagrams from two sources: The conceptual design phase developed preliminary parameters at the subsystems level and above; and the data development deliverable produced item-level parameters.

The allocated parameters will be subject to revision because the tasks of the later deliverables will test them against additional design detail. The possible iterations are as follows:

Task legend:
AB3.4/5 Develop reliability, maintainability, & life cycle cost data.
AB3.7 Define system for continual collection and processing of data.
AB8.7 Rank the consequences of each failure.
AB10.8 Determine candidate availability schemes and refinements.
AB11.5 Determine solutions for plant layout constraints to maintainability.
AD1.5 Review & revise allocated availability parameters.

Deliverables legend:
TB4 Performance specifications and select & size equipment.
TD2 Detailed safety & hazards analysis.

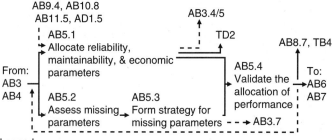

Legend:
- - ► = Iteration task
AB = Availability deliverable or task in the basic design phase.
AD = Availability deliverable or task in the detailed design phase.
TB = Traditional deliverable or task in the basic design phase.
TD = Traditional deliverable or task in the detailed design phase.
AE&M = Availability engineering & management

Fig. 6–6 Tasks to allocate availability parameters to plant logic diagrams (AB5 of Figure 6-1).

- Maintenance logic-tree analysis (AB9, Chapter 8) will determine maintenance strategies for each identified significant failure. These will ultimately be applied to validate the allocated availability parameters (AB9.4). The selected strategies can affect reliability, maintainability, and economics.
- The availability scheme is eventually optimized (AB10, Chapter 9). In that process, alternate candidate schemes are formulated (AB10.8) that may generate requirements for alternate allocations.
- The plant layout is later assessed for maintainability (AB11). It is tested against the maintenance actions selected for significant failures. Accordingly, maintainability and economic parameters may require revisions (AB11.5). This is because the plant layout may have constraining conditions.
- The maintenance tasks analysis in the detailed design phase will carefully detail each maintenance task. Its result will be refinements in time-to-maintain and costs (AD1.5, Chapter 11).

The allocated parameters will flow to other deliverables as follows:

- There may be iterations to the earlier tasks to develop data for availability design (AB3.4 and AB3.5, Chapter 6).
- The parameters will also flow to the safety hazards analysis deliverable of the traditional detailed design (TD2, Chapter 14). The parameters enable the analysis to be quantitative and provide frequency of occurrence data.

AB5.2. Assess importance of missing and poorly-developed parameters. Some parameters may be missing or poorly developed because of the poor condition or absence of data. Also, it may not be economically feasible to develop some data. Therefore, this task is to assess the significance of the data to overall availability performance. This is done with preliminary availability and economic analyses models.

AB5.3. Formulate strategy for problem parameters. It was mentioned that allowances must be made for missing parameters. Thus, this task is to establish the approaches to those allowances that involve a process of assumptions, analysis, and decisions. The strategies should also reflect the vision of life-cycle availability management rather than just the initial design. Figure 6-6 shows that these strategies will be made part of the design for the life-cycle collection and management of data (AB3.7, Chapter 6).

AB5.4. Validate the allocation of performance.   The allocation of reliability and maintainability parameters to the plant logic diagrams implies an expectation of availability performance. This task is to verify that availability performance is consistent with management's specified plant performance.[3]

The conceptual design phase quantified plant-level availability (AC2). It then translated the determination to preliminary parameters at the subsystem level. Consequently, the preliminary availability model can now be used to confirm that the element-level allocations produce the subsystem level parameters. Ultimately, the two levels of parameters will be reconciled to produce the specified plant performance.

Figure 6-6 shows that the results are inputs to tasks in two lateral deliverables and traditional designs. The cases are as follows:

- The consequences of identified failures are determined (AB8.7) as part of FMECA (AB8, Chapter 8). The final allocation of parameters will provide information with respect to the frequency of a failure.
- The result of this task is also input to traditional equipment performance specifications, sizing, and selection (TB4, Chapter 10).

Allocation Compared to Optimization of Parameters

Availability parameters have now been allocated to the plant logic diagrams. However, this model does not qualify as the plant availability scheme. This is because the plant is a complex system of parameters. Computer-based models are necessary to fully understand and optimize the now massive detail.

Dealing with this massive detail is the purpose of the availability and economic models (AB6 and AB7, Chapter 7). The parameters will be further analyzed and developed by the the subsequent availability analysis (AB10, Chapter 9). The balance of reliability and maintainability with that plant performance is optimized.

The Production Process and Availability Parameters Compared

It was mentioned earlier that the plant's production process and its optimal performance is a network of process parameters. In a chemical process plant, for example, these parameters include temperature, flow rate, compo-

---

[3] Over time, plant performance is a moving target. The better the plant performs, the better management will want it to perform. This spiral of expectations has immense ramifications for the owner's business results.

sition, pressure, etc. The parameters must be identified for each function in the process flow diagrams.

This is equivalent to the synthesis of plant logic diagrams and availability parameters. It also demonstrates why traditional plant design can only center on the production process and its necessary equipment. By comparison, availability performance is a very different problem. It uses different tools and deliverables to optimize a different fundamental set of plant performance parameters.

Without these processes, availability design can only be based on intuition. At best, availability analysis is done coarsely at the top level. It is obvious that the design for availability performance is not within the human capacity to solve by intuition. Nor will coarse design methods be effective. By comparison, nobody would consider designing the production process with the same minimal rigor.

Bibliography

Blanchard, Benjamin S. *Logistics Engineering and Management.* 4th ed. Englewood Cliffs, N.J., Prentice Hall, 1991.

Blanchard, Benjamin S., and Fabryky, Wolter J. *Systems Engineering and Analysis.* 2nd ed. Englewood Cliffs, N.J., Prentice Hall, 1990.

Bloch, Hienz P. and Geitner, Fred K. *An Introduction to Machinery Reliability Assessment.* New York, Van Nostrand Reinhold, 1990.

Bloch, Hienz P. *Practical Machinery Management for Process Plants.* Volume 1. *Improving Machinery Reliability.* 2nd ed. Houston, Tx., Gulf Publishing Company, 1988.

Cleveland, E.B., Jarrett, A.A., and Morrison, C.B. Assessment of Methods for Implementing Availability Engineering in Electrical Power Plants. For the Electric Power Research Institute by Holmes & Narver, Inc., May 1977. EPRI NP-493.

Electric Power Research Institute. Demonstration of Reliability Centered Maintenance. Volume 1. Project Description. Palo Alto, Calif., January 1989. EPRI NP-6152.

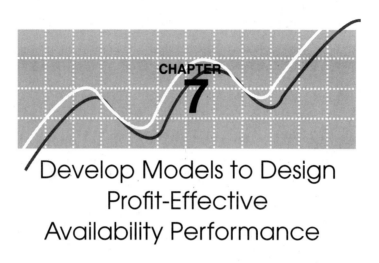

CHAPTER

7

Develop Models to Design Profit-Effective Availability Performance

Introduction

Chapters 5 and 6 described two requirements of the basic design phase. One established the infrastructure for the design, improvement, and management of availability performance. The other detailed the plant as a system of reliability, maintainability, and economic parameters.

The next step is to develop the models that will be used to evaluate and optimize availability performance. The plant logic diagrams determine their structure. The formulated parameters are their input. The models will ultimately reside in the life-cycle improvement, change, and data management systems.

The deliverables for developing the models are shown in Figure 7-1. Accordingly, it also shows that this chapter will describe the following:

- The model for computing plant availability from the expected reliability and maintainability performance of its elements (AB6).
- The model for financially evaluating the availability scheme (AB7).

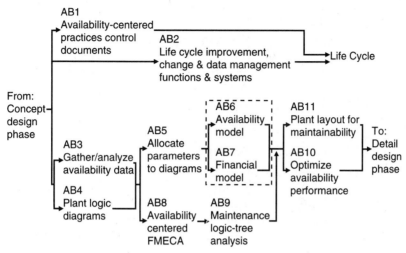

AB = Availability deliverable in Basic design phase
FMECA = Failure modes, effects & criticality analysis

Fig. 7–1 Basic design phase availability engineering deliverables for developing availability.

AB6. Develop the Plant Availability Model

Definition and Purpose

The availability model calculates plant availability performance. The model's program mathematically replicates the plant logic diagrams. Its outcome is a probability distribution.

Why the Model is Necessary

The system of allocated parameters from the previous deliverables was classified as preliminary for a simple reason. There are now many elements in the plant logic diagrams. Each has been assigned reliability and maintainability parameters, and each is expressed as a probability distribution rather than a cardinal number.

These parameters must still be related, tested, and aligned as an optimal set of parameters. The calculation is iterative in search of that optimum. Furthermore, the availability calculation is done many times for a single answer.

The plant availability model is the tool that serves that requirement. It is necessary because the calculations are too complex, cumbersome, and iterative for manual approaches. Without it, the capacity to design and manage performance is limited.

Concept and Approach of the Model

Three fundamental formulas. The availability model essentially is calculated from the inside out. Three formulas are involved. The discussion that follows will describe the process in a simple application. The objective is to describe the availability modeling concept and approach.

The first fundamental formula is the calculation of availability. It is applied to each element in the plant logic diagram. The calculation is as follows:

$$A = \frac{TBF}{TBF + TTM} \qquad \text{(7-1)}$$

Where:　A = Availability as a ratio or percentage of time.
　　　TBF = Time-between-failure.
　　　TTM = Time-to-maintain.

Notice that the formula does not show mean time-to-failure and mean time-to-maintain.

The second and third formulas are probability formulas. They reflect plant elements related to each other either in parallel or in series. They are as follows for elements "A" and "B".

- For independent elements in series:

$$A = (A_A) (A_B) \qquad \text{(7-2)}$$

- For independent elements in parallel:

$$A = 1 - (1 - A_A) (1 - A_B) \qquad \text{(7-3)}$$

There are other relationships between equipment, assemblies and components. For example, the failure of an item may not be an independent event. It may be a consequence of another failure. The model builders (mathematicians, statisticians, reliability engineers, etc.) can apply different probability equations to fit any type of interrelationship.

The calculation of system availability. As mentioned, the model essentially computes availability from the inside out. Figure 7-2 provides a simple

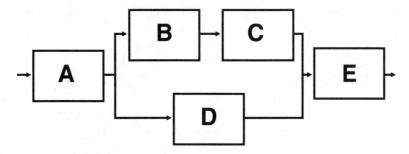

Fig. 7–2 A simple example of a logic diagram to demonstrate the calculation of availability.

plant logic diagram to demonstrate the process. Its elements are "A" through "E". The steps are as follows:

1. Compute availability for each element in the plant logic diagram. This is done with Equation 7-1.
2. Starting from the inside, "B" and "C" are elements in series. Equation 7-2 for elements in series is applied as follows:

$$A_{BC} = (A_B) (A_C) \tag{7-4}$$

3. The results of Equation 7-4 and element "D" are now parallel elements. Equation 7-3 for elements in parallel is applied as follows:

$$A_{BCD} = 1 - (1 - A_{BC}) (1 - A_D) \tag{7-5}$$

4. Finally, a series relationships remains. Their combined availability is computed with Equation 7-2 as follows:

$$A = (A_A) (1 - A_{BCD}) (A_E) \tag{7-6}$$

Plant availability distribution by Monte Carlo simulation. The previously allocated reliability and maintainability parameters can be expressed as probability distribution functions. The objective is to compute plant-level availability as a probability distribution.

The common method for computing probability distributions is the Monte Carlo simulation. It can be used to generate the necessary plant-level expectation of availability performance.

There are also mathematical methods to calculate the probability distribution, but they are more costly. The cost and time expended to develop the calculation is not justifiable in lieu of what can be achieved with Monte Carlo simulation.

Monte Carlo simulation uses random number methods to sample the reliability and maintainability distributions of each plant element. This is done with random number methods. This realistic sequence of artificial events provides a means of viewing a system's performance. The quality of the result depends on the quality of the sampled probability distributions.

It was noted that Equation 7-1 reflected "time" rather than "mean time." This is because the simulation process randomly selects time intervals for reliability (time-between-failure) and maintainability (time-to-maintain) for each plant element. These are taken from the respective distributions for each element. The calculation process is then executed for each element. It is computed hundreds of times to generate a distribution of availability performance.

Tasks for Accomplishing the Deliverable of Developing the Availability Model

The tasks for developing the availability model are described in the next section. They are flowcharted in Figure 7-3. The flowchart also shows the sequences and iterations with tasks in other deliverables.

AB6.1. Analyze model uses over plant life.   The model will first be used in initial plant design. However, its use will continue over the plant's life in the management and improvement of plant availability performance.

Accordingly, this task is to identify and analyze the following with respect to the owner's organization and its business processes:

- All possible uses of the model throughout the plant's life.
- The performance and management purpose of each use.
- The skills and expertise of potential users.
- The means of access and utilization by each user.
- The need for incorporating the model into other corporate, operating company, and plant-level design, analysis, and management systems. This includes the need to draw upon and provide input to associated databases.

An earlier task determined which organizational functions will affect the top-level factors of availability (AB1.3, Chapter 5). The search was part of the deliverable for developing availability-centered practices documents. Figure 7-3 shows that its findings provide important input to this task.

AB6.2. Identify stages of model development. The availability performance model is not a static development. The quality of its computation evolves along with plant availability experience and data. Preparing for this possibility requires the following deliberations:

- What are the anticipated stages in the plant's life that will allow a better quality forecast of availability performance? An example is to identify when important data will reach statistical milestones.
- How will the availability model be changed in response to the stages and milestones? The formulas applied to model elements may need to be changed as the statistical quality of the data is improved.
- What other availability engineering and management processes must be evaluated and changed? For example, the data collection and analysis processes may be changed to reflect improving data.

AB6.3. Identify the candidate models. Candidate models must be identified and screened. Availability models are now widely available. They range from inexpensive desktop systems to very powerful ones. An additional source is custom designed models. In the middle is a combinations of these.

The powerful models are often available through specialized consultants who have developed their own systems. The choice between them is a function of their comparative abilities. The choice may also reflect the consultant's ability and willingness to customize its baseline version of the model.

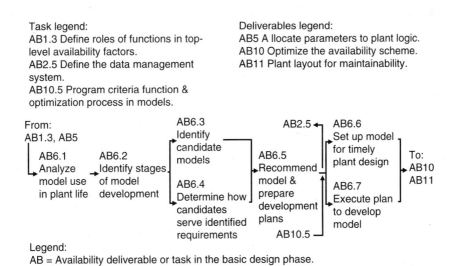

Task legend:
AB1.3 Define roles of functions in top-level availability factors.
AB2.5 Define the data management system.
AB10.5 Program criteria function & optimization process in models.

Deliverables legend:
AB5 Allocate parameters to plant logic.
AB10 Optimize the availability scheme.
AB11 Plant layout for maintainability.

From:
AB1.3, AB5

AB6.1 Analyze model use in plant life

AB6.2 Identify stages of model development

AB6.3 Identify candidate models

AB6.4 Determine how candidates serve identified requirements

AB6.5 Recommend model & prepare development plans

AB2.5

AB10.5

AB6.6 Set up model for timely plant design

AB6.7 Execute plan to develop model

To:
AB10
AB11

Legend:
AB = Availability deliverable or task in the basic design phase.

Fig. 7–3 Tasks for developing the plant availability model (AB6 of Figure 7-1).

AB6.4. Determine how each candidate model can serve identified requirements.   Previous tasks identified what is needed for plant design and subsequent availability management. Candidate models and their providers must be evaluated against these requirements.

A fundamental issue is whether a candidate can be customized when necessary. Thus, one purpose of the task is to compare systems with respect to required modifications. This assessment is made in terms of the need and capacity for modifications.

AB6.5. Make recommendations and prepare model development plans.
A model and source must ultimately be recommended. A plan must also be formulated for the model's life-cycle development. This should include the work scope, schedule, and costs for its final detailed definition, physical design, and implementation. The plans must then be integrated with the requirements for the financial model (AB7.7).

This task is carefully integrated with the plans to develop the availability-centered improvement, change, and data management systems and functions (AB2). Consequently, its results flow to the task of defining the life-cycle data management system (AB2.5, Chapter 5). The purpose is to incorporate the model in the overall data management functions and systems.

AB6.6. Set up the model for timely plant design.   The methodical development of the model may not be timely enough to serve the immediate needs of the project. The calculations of the model must be available to monitor the traditional design deliverables and tasks. Thus, this task is to program and implement the model to serve the project's immediate needs.

AB6.7. Execute the plan to develop the model.   This and the previous task are part of implementing the plan for the detailed definition, physical design, and implementation of the model. The preceding task was concerned only with serving the needs of the design process. This task is concerned with the model's full scale development. Both goals must be integrated into a single plan.

Figure 7-3 shows that this and the preceding task receive input from the deliverable of analyzing and optimizing the plant availability scheme (AB10, Chapter 9). Another task will have formulated two model components to be incorporated in this and the preceding task (AB10.5). The components are:

- A criteria function developed to value and compare alternate availability schemes.
- The mechanisms or process for determining an optimal scheme.

AB7. Develop the Plant Financial Model

Definition and Purpose

The availability model solves the first fundamental objective of availability engineering and management. To achieve a specified availability performance, it connects reliability and maintainability by design rather than by intuition and default. The second equally important fundamental objective is to determine the most financially attractive life-cycle solution for a specified level of performance. Applied along with the availability model, the financial model is the tool that serves this objective.

The Analysis, Design, and Management of Financial Strategies

Financial analysis goes beyond cash flow and net present value analysis. Furthermore, plant economics are not always simply the maximization of profits and productivity of capital working assets. There may be other strategic issues. For example, an owner may wish to act as if capital poor and working asset rich. There may be debt levels and other financial limits. In other words, the project may be confronted by capital constraints to be offset with working assets solutions.

These situations may lead to paying for productive capacity by incurring costs during the plant's functioning life instead of more capital intensive approaches. A plant with high reliability from configuration and equipment selection is an example of the latter. By comparison, the owner can build a plant with less sparing of equipment, less durable equipment, etc. The objective is still to create a plant that can perform its intended availability in a profitable manner.

Philosophy of the Model

Basic form.   Plant life-cycle cost and revenue forecasts should be constructed in spreadsheet form. The format should parallel the owner's normal financial statements and reports. However, its detail is not limited to assets, expenses, and revenues associated with availability performance.

A spreadsheet philosophy allows a view of the plant's life cycle. The cause and effect relationships of availability decisions between accounts and line items can be traced and studied. This is also the case for cause and effect from one financial period to the next.

Cross-organizational participation.   Corporate, operating company and plant management functions should participate in the model's development. They represent its various financial elements. This is especially so if management wishes to develop a model for all dimensions of plant performance. The availability design team should explore this possibility early in the project planning stage.

Financial Statement Subsystems of the Model

Structuring the model to parallel the owner's financial statements involves the following subsystems.

- The income statement.
- The balance sheet.
- The sources and uses of working assets.
- Life-cycle investment and the activity cost subsystem.

They are described in the next sections as individual subsystems. However, they are integrated because the products of each depend upon the others.

Income statement subsystem.   The classic goal of financial accounting is to measure current and cumulative historical performance. Availability engineering and management extends this purpose to the future. Accordingly, the model's income statement subsystem serves the following purposes in life-cycle analysis, design, and management:

- Matching plant investment to the short- and long-term availability performance it creates. Investment appears as depreciation expense.
- Matching availability management and maintenance operations costs to that availability performance.
- Optimizing the interrelationship of investment, functioning cost, and revenues. The goal is maximum life-cycle income.
- Assessing a design for its operating leverage. Operating leverage is the analysis of two dimensions: the percentage of change in income that results from a change in availability; and the percentage of change in income that results from a change in availability-centered expenses.
- Providing partial information for designing and managing availability performance for maximum return on investment. Income as a percentage of revenues is one of two factors in this return.

These functions place management and its design team in a position to explore the income-centered significance of design, improvement, and management decisions. This exploration is with respect to income margin, timing, and lifetime total.

Balance sheet subsystem.   The balance sheet subsystem explores the affect of design and management decisions on capital and working assets. This is important for the following reasons:

- They provide choices for working-asset versus capital-asset approaches to a specified availability performance. The balance sheet subsystem is the tool for designing across these choices.
- The subsystem serves as a partial source of information for designing for maximum return on investment. Productivity of working and capital assets (revenue divided by assets) is one factor of the computation. As mentioned, the other factor is income as a percentage of revenue.[1]

Sources and uses of the working assets subsystem.   The financial statement includes what many know as a statement of sources and uses of working assets. This is the cash-flow subsystem of the financial statements. Its purposes are as follows:

- To provide a basis for computing time-to-payback, net present value, and internal rate of return.
- To enable the analysis and design of availability performance with regard to working assets, liquidity, and debt capacity.

Life-cycle investment and activity cost subsystem.   The above subsystems draw from the life-cycle investment and activity cost subsystem. Life-cycle costs were first determined as part of the deliverable to develop availability data (AB3). One of its tasks (AB3.5) detailed each item for its capital and direct maintenance costs. Those results flow to this subsystem. They are later refined by tasks in the detailed design phase.

Thus, the subsystem functions as is a clearing house of data for the preceding subsystems. Investment data flows to the balance sheet subsystem. Direct maintenance expenses flow to the income statement subsystem. They are combined there with the many indirect or overhead expenses of availability performance. The indirect expenses are developed from other sources in the detailed design phase.

---

[1] Return on investment equals income as a percentage of revenues times revenues divided by assets.

Readout Meters of Availability Performance

The financial model is constructed with "readout meters" of performance. These are used to initially evaluate and optimize plant availability design and improvement. Just as important, they are used to determine the appropriate benchmarks for normal plant performance.

There are three categories of readout measures:

- Key financial statement items.
- Ratios for line items and balances.
- Graphical readouts.

Key financial statement items.   Readouts include key items from the financial statements. These include revenue (from volume), income, expenses, and asset balances. Usually, these are the fundamental items that management most frequently monitors performance against. They are also often the measures applied to management by outside entities.

Ratios for line items and balances.   Ratios relate financial statement account balances and line items for analysis. They may involve items within the same statement or items in separate statements.

The ratios test current position, equity position, and operating results. As readouts to availability performance, their nature and expected importance are as follows:

- Ratios for the current position relate working capital to liquidity. They are most relevant to the decisions to be made for debt and liability. These decisions are not generally within the domain of availability engineering and management: They typically belong to the domain of business management. Therefore, these ratios are not greatly relevant to an availability-centered financial model. However, it is conceivable that they could be of some relevance.
- Ratios of equity position show financial consequences of availability design choices between capital versus working assets. For this reason, they are of interest to availability design and management. They can also be used in the design process to measure longer-term capacity for generating or acquiring investment resources.
- Ratios to measure operating results are the most important to availability engineering and management. They provide measures of income and productivity of assets. More importantly, they are used to develop a balance between them. These ratios are packaged in the Du Pont Return on Investment Model.

Graphical readouts. The financial model is also built to generate graphical presentations of life-cycle financial performance. They provide life-cycle profiles of key financial account balances, line items, and ratios.

Graphical readouts are important because single calculations, income and costs figures, and tables can easily disguise important consequences.

Tasks for Accomplishing the Deliverable of Developing the Plant Financial Model for Availability Performance

The basic tasks for developing the availability-centered financial model are described in the following sections. They are flowcharted in Figure 7-4. The flowchart also shows the sequences and iterations with tasks in parallel and previous deliverables. Furthermore, the tasks are closely integrated with the design and development of the availability model (AB6).

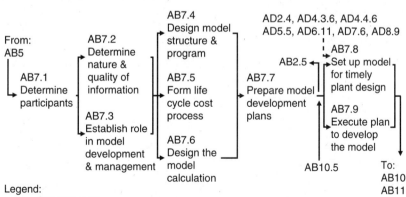

Task legend:
AB2.5 Define the data management system.
AB10.5 Program criteria function & optimization process in models.
AD2.4 Estimate costs of technical materials operation.
AD4.3.6 Determine confidence & provision levels & costs of service resources.
AD4.4.6 Determine stocking levels & costs of consumable resources.
AD5.5 Cost analysis of the facility scheme.

AD6.11 Cost analysis of the training program.
AD7.6 Estimate costs of maintenance operation functions.
AD8.9 Determine total cost of availability organization.

Deliverables legend:
AB5 Allocate parameters plant logic.
AB10 Optimize the availabiality scheme.
AB11 Plant layout for maintainability.

Legend:
- - ► = Iteration task
AB = Availability deliverable or task in the basic design phase.
AD = Availability deliverable or task in the detailed design phase.

Fig. 7–4 Tasks for developing the plant financial model (AB7 of Figure 7-1).

AB7.1. Determine participants.   Who participates in the model's development reflects which organization functions provide business input to the model or use its results. This is because the financial model is a collaboration of corporate, operating company, and plant entities. Some examples of the these are the functions responsible for strategic planning, long-term financial planning, operating budgets, accounting, etc.

Who participates may also reveal existing financial formats, models, and systems. The financial model may copy them, incorporate them, or access their databases.

AB7.2. Determine the nature and quality of information.   Information needs will be determined inthis task. The concern is what is needed, why, and when. The quality of the information and the form of delivery must also be established.

AB7.3. Establish roles in model development and management.   This task is to establish the initial design and subsequent management roles, responsibilities, and authorities of the participants. They are formulated as a preliminary management structure.

This and the preceding tasks are ultimately integrated with the deliverable to develop improvement, change, and data management functions and systems (AB2).

AB7.4. Design the model structure and program.   The content of the model must be defined in detail. This definition includes the following:

- The model's spreadsheet dimensions and their detail. One dimension is the life cycle and its time intervals. The other is financial line items and account balances.

  The second dimension has multiple categories and possibilities. Some of these are: financial statement subsystems, categories of life-cycle costs, and performance measures and financial profiles.

- Contents of underlying backup detail. These are a function of information included in the model and gathered from its uses. A plant-level example is the detail gathered from life-cycle cost analysis.(See AB7.5).
- Relationships between account balances, line items, and cost categories.
- Relationships to be created as "readout meters."
- Model inputs and outputs. These identify suppliers, customers, and the owners of the data.

AB7.5. Formulate the life-cycle cost process. Life-cycle costs are a major component of the model. Accordingly, this task is to develop the model's life-cycle cost subsystem. The approach is as follows:

- Establish the life-cycle phases of costs. These are typically design, construction, startup, producing life, and retirement. It is conceivable that these may be subdivided. For example, there may be known development stages expected to occur in the future.
- Meticulously enumerate the cost items in each phase. This provides a checklist for individual cost analysts.
- Establish the cost relationships and profiles for each cost item. The relationship of each item to the life-cycle and between costs must be modeled. This capability directly serves the task of selecting maintenance strategies for each failure (AB9.2, Chapter 8) because some strategies will be selected based on economics.
- Establish cost reporting formats.

As mentioned, life-cycle costs were first developed by a task in the deliverable of gathering and analyzing data (AB3.5, Chapter 6). Thus, the first two steps may have already been accomplished. If so, they may be refined by this task.

AB7.6. Design the model calculation. The model's integrated system of calculations must be designed. Therefore, this task is to formulate these calculations. The subject relationships were rigorously defined by the preceding tasks.

AB7.7. Prepare model development plans. A plan must be formulated for the model's life-cycle development. This includes schedules and costs for initial detailed definition, physical design, and implementation. The plans must then be integrated with the same requirement for the availability model (AB6.5).

Figure 7-4 shows that the results will flow to the task of defining the life-cycle data management system (AB2.5, Chapter 5). The purpose is to incorporate the model in the overall availability-centered data management functions and systems.

AB7.8. Set up the model for timely plant design. The methodical development process may not be timely enough to serve the immediate plant project needs. Therefore, the model's planned calculation must be made available to monitor the traditional design deliverables and tasks. Thus, this task concentrates on serving immediate modeling needs.

Figure 7-4 shows that the model will be refined by evolving input from later deliverables. Their work products will variously flow into the income statement, balance sheet, and life-cycle cost subsystems. These inputs are:

- The technical materials operation (AD2.4, Chapter 11).
- Provisioning of human, material, and facility resources (AD4.3.6, AD4.4.6, and AD5.5, Chapter 12).
- Development of the training program (AD6.11, Chapter 12).
- The costs of the maintenance operation and overall availability organization (AD7.6 and AD8.9, Chapter 13).

The figure does not show the direct labor and material cost of each maintenance task. They flow into the model via the availability parameters allocated to plant logic diagrams. They are actually initial estimates that will be refined by maintenance task analysis in the detailed design phase.

AB7.9. Execute the plan to develop the model.   This and the preceding task will implement the plan for the detailed definition, physical design, and implementation of the model. The preceding task, however, was concerned with serving the progressing plant design. This task is concerned with the model's full-scale development.

Figure 7-4 shows that this and the preceding task receive input from the deliverable for optimizing the plant availability scheme (AB10, Chapter 9). Another task will have formulated two model components to be incorporated in these tasks (AB10.5). They are as follows:

- A criteria function developed to value and compare alternate availability schemes.
- The mechanisms or processes for determining an optimal scheme.

Limitations of Cost Accounting

The financial model of availability performance is important to the quest for accountability. Maintenance operations, as part of availability management, are a primary cost of business. However, normal cost accounting practices make it difficult to the point of being impractical to account for their full costs and associated production results. This is because many of the costs are lost in overhead accounts.

The financial model does not suffer from this weakness. Instead, it allows the availability engineering and management processes to analyze all costs and consequences of plant availability performance.

Bibliography

Blanchard, Benjamin S. *Logistics Engineering and Management*. 4th ed. Englewood Cliffs, N.J., Prentice Hall, 1991.

Blanchard, Benjamin S., and Fabryky, Wolter J. *Systems Engineering and Analysis*. 2nd ed. Englewood Cliffs, N.J., Prentice Hall, 1990.

Davis, Gordon B. Management Information Systems: *Conceptual Foundations, Structure and Development*. 2nd ed. New York, McGraw Hill, 1985.

Fabryky, Wolter J. and Blanchard, Benjamin S. *Life Cycle Cost and Economic Analysis*. Englewood Cliffs, N.J., Prentice Hall, 1991.

Horngren, Charles T., Foster, George and Srikant, M. Datar. *Cost Accounting, A Managerial Emphasis*. 8th ed. Englewood Cliffs, N.J., Prentice Hall, 1994.

Kieso, Donald E. and Waygandt, Jerry J. *Intermediate Accounting*. 7th ed. New York, John Wiley & Sons, 1991.

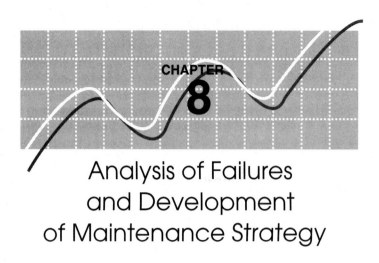

8

Analysis of Failures and Development of Maintenance Strategy

Introduction

The previous two chapters described how to develop data and information, to formulate them as parameters to depict the plant as a system, and to create models to analyze and optimize that system. This chapter introduces another set of requirements. One is to identify and analyze all significant failures in the plant. The other is to determine feasible maintenance strategies for each.

The deliverables for assessing failures and formulating strategies are shown in Figure 8-1. Accordingly, it is shown that two deliverables are the subject of this chapter. They are:

- Availability-centered failure modes, effects, and criticality analysis (AB8). This deliverable is to identify and analyze each significant failure.
- Maintenance logic-tree analysis (AB9) for determining a maintenance strategy for each identified failure.

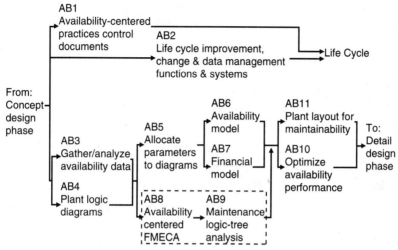

AB = Availability deliverable in Basic design phase
FMECA = Failure modes, effects & criticality analysis

Fig. 8–1 Basic design phase availability engineering deliverables for the analysis of failures and development of maintenance strategy.

AB8. Availability-Centered Failure Modes, Effects, and Criticality Analysis

Definition and Purpose

A fundamental part of availability performance is designing and planning for significant failures. Failure modes, effects, and criticality analysis (FMECA) serve that need. It is a tool for identifying failures and assessing the criticality of their consequences. The results will determine how to prevent significant failures that can be avoided and to limit the effect of those that cannot be avoided.

Proactive Problem Solving

Long after design, plants frequently require problem-solving initiatives. For single problems, these initiatives intermittently and fully involve from few to many people.

The challenge is usually substantial. A team must try to unravel a chain of events leading to the problem. Just isolating the problem's boundary and elements can be difficult.

The challenge is made even more difficult by timing. The team does not have the same holistic knowledge that the original design team once had.

Therefore, FMECA is often viewed as a proactive problem-solving tool because it often solves problems that could have been solved during the plant's design phases.

A proactive solution process is substantially more efficient and effective. In turn, the plant maintenance operations can be formed to manage, rather than be surprised by, these problems.

Tasks for Accomplishing the Deliverable for Failure Modes, Effects, and Criticality Analysis (FMECA)

The tasks for FMECA as a deliverable are described in the following sections. Their sequence is flowcharted in Figure 8-2. The flowchart also shows the sequences and iterations with tasks in parallel and previous deliverables.

It is not the intent of this chapter to provide an intricately detailed description of the FMECA steps. Literature that details the exact steps is widely available. Many books even include forms to be completed during analysis. Furthermore, there are many variations to choose from. Thus, to attempt such detail is misleading since there is more than one possible approach.

Therefore, the following deliverables describe FMECA as a set of fundamental project tasks in the design of availability performance.

AB8.1. Select a FMECA method.   As mentioned, there are many methods of failure modes analysis. The most common is the tabular method. Fault-tree analysis is another form. The point is that a method must be selected to fit the subject project, organization, and other initiatives. That is the purpose of this task.

AB8.2. Define and set up the FMECA and maintenance logic-tree analysis system.   FMECA and subsequent maintenance logic-tree analyses are part of the plant's living availability design.

This task is to define the system to allow the capture and later access to both methods and their evolving findings. It is also to plan to set up the defined system as an essential element for serving immediate project needs.

The results will be incorporated in the definition of the life-cycle data management system (AB2.5, Chapter 5). This is shown in Figure 8-2.

AB8.3. Identify all failures.   FMECA begins by identifying all significant possible failure modes. A failure mode may be "fails to start," "fails to

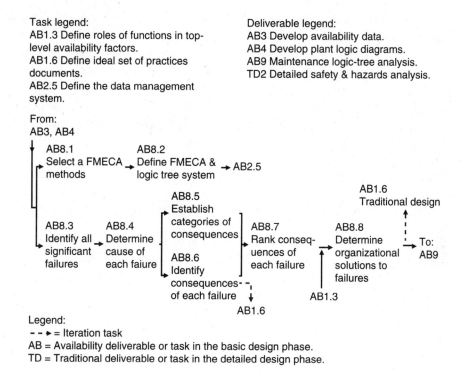

Task legend:
AB1.3 Define roles of functions in top-
level availability factors.
AB1.6 Define ideal set of practices
documents.
AB2.5 Define the data management
system.

Deliverable legend:
AB3 Develop availability data.
AB4 Develop plant logic diagrams.
AB9 Maintenance logic-tree analysis.
TD2 Detailed safety & hazards analysis.

From:
AB3, AB4

AB8.1
Select a FMECA
methods

AB8.2
Define FMECA &
logic tree system

→ AB2.5

AB8.3
Identify all
significant
failures

AB8.4
Determine
cause of
each faiure

AB8.5
Establish
categories of
consequences

AB8.6
Identify
consequences
of each failure

AB8.7
Rank conseq-
uences of
each failure

AB8.8
Determine
organizational
solutions to
failures

AB1.6
Traditional design

To:
AB9

AB1.3

AB1.6

Legend:
- - ► = Iteration task
AB = Availability deliverable or task in the basic design phase.
TD = Traditional deliverable or task in the detailed design phase.

Fig. 8–2 Tasks for developing failure modes, effects, and criticality analysis (AB8 of Figure 8-1)

achieve pressure," etc. They are identified to the levels at which specific design, procurement, construction and startup, operations, or maintenance solutions can be developed.

AB8.4. Determine the cause of each failure. The next task is to determine the cause of each failure. Its objective is to trace the sequence of initiating and interim events leading to the failure.

The search will span the plant's life cycle instead of searching only for those failures that occur during plant functioning. Accordingly, causes originating in the following life cycle phases are identified:

- The design phases: These failures are a result of decisions concerning configuration, service levels, materials, selection among types of equipment, etc.
- The manufacturing phase: Failures during this phase are not just from flawed production and fabrication. They are caused by the failure to discover problems during inspection and testing. These are failures of

the manufacturer's, project contractor's, and owner's quality assurance and control programs.

- The construction and startup phase: Problems during this phase result from the failure to protect equipment, faulty installation, failure to properly test and evaluate performance, poor initial startup procedures, etc.
- The commercial production phase: Failures during this phase result from operating and maintenance practices and conditions.

AB8.5. Establish categories of consequences. The study team should establish the categories of consequences. This is actually in the domain of management policy. Thus, it was first broached in the conceptual design phase (AC1.6, Chapter 3).

Consequences can go beyond just availability related consequences. They may also include issues associated with various safety codes and regulations, loss of productivity, etc. In fact, all failures have a safety or economic consequence.

Accordingly, FMECA may recognize the following categories of critical consequences:

- Strategic and tactical consequences. These focus on suppliers and markets.
- Loss of plant availability. This is to be shut down or forced to operate at a lower production level.
- Increased exposure to the loss of availability. As items fail, the plant does not always experience a loss of availability or a lower production level. However, the mean of expected plant availability shifts to the left. Its distribution takes on a less attractive shape. This case was shown in Figure 1-3. As a result, expected performance is less than desired.
- Risks to people, equipment, and the environment.
- Loss of production-cycle productivity. These included increased wastes and the costs of their disposal. These may be absorbed by the regional environment and economy.

AB8.6. Identify the consequences of each failure. This task is to assess the consequences of each failure. It is a challenge because the ultimate consequences are not always just the next event. Instead, there may be subtle and far-reaching results. Various cause and effect analysis methods can be used to make this determination.

The identified consequences will provide verifying information for the development of availability-centered practices documents (AB1, Chapter 5). An ideal set of practices has been defined already (AB1.6). Since this task is to identify critical design and management aspects, it is an opportunity to check and refine that determination.

AB8.7. Rank the consequences of each failure. This task is to rank the consequences of each failure according to its category. One method is to develop a multidimensional index, which is a composite of the following factors:

- The severity of the consequences.
- The probability of their occurrence.
- The ability to detect the failure in advance.

Severity ranking must be developed as a classification system for the subject plant. An example of ranking is as follows:

1. Results in a death or a loss of the system.
2. Results in an injury or in the loss of the system to short- and longer-term production cycles.
3. Results in a minor injury or in the damage of a subsystem. The latter will reduce the plant's capacity to produce.
4. Significantly reduces the short-term probability of experiencing the specified availability.

The example shows that there are two types of availability: One is normal loss or reduction of availability or the expected failures and planned shutdowns during normal functioning.

The other is the abnormal loss of the physical availability to perform. This is a catastrophic event because the plant may no longer be able to produce, or it is able to produce only at a diminished capacity. The consequences of these events often affect the plant beyond the immediate term, since some physical aspect of the plant is damaged.

These consequences are in the domain of production process safety management. However, they are served by availability engineering and management with respect to mechanical integrity. This is a normal part of the discipline's overall process.

Probability of occurrence is the second component of the composite index. A failure of high severity but rare occurrence is very different than one with a high rate of occurrence. The feasible solutions for eliminating or controlling them may be different. Figure 8-2 shows that probability data is provided by the final task of allocating availability parameters to plant logic diagrams (AB5.5, Chapter 6).

The third component of the index is the ability to detect and remedy a pending failure. It is a valuable advantage because although the consequences cannot be reduced, their likelihood can be limited. Highly signifi-cant failures that cannot be detected are very important to control because they may have grave consequences.

The three factors are then combined in a critically index. However, this may create new peril. An example is a severe failure that occurs infrequently. It is still a severe failure. An index could hide that severity by the virtue of its calculation. The result is a false sense of security that may increase the plant's exposure to the very consequence management most wishes to avoid.

Figure 8-2 shows that the determined consequences are also important to detailed design phase risk analysis (TD2, Chapter 14) in the detailed design phase. Since it is a traditional deliverable, the results of this task flow to whatever methods the safety team elects to apply in the assessment of risk.

AB8.8. Determine organizational solutions to failures.   This task is to identify organization level solutions for each failure. Special attention is given to the most critical. There are four possible solutions:

- Design
- Operation
- Maintenance
- Quality assurance

In an earlier deliverable, the roles of organization functions were identi-fied with respect to top-level factors of availability (AB1.3, Chapter 5). Figure 8-2 shows that the findings provide information for this task that will help to identify possible organizational solutions.

The choice between four possible solutions depends on the progress of the plant's design and acquisition. It is preferable to make design changes early in the design phase because it is easier to do at that time. Figure 8-2 shows that this choice will iterate with the traditional design deliverables. The objective is to eliminate consequences. Possibilities include the follow-ing:

- Eliminating the failure mode.
- Reducing the failure rate.
- Reducing the consequences. This and the previous options are accom-plished by modifying components, eliminating a component, or modi-fying the system.

As the design progresses and becomes increasingly more rigid, operat-ing and maintenance solutions will become more likely. Maintenance solu-tions are the focus of the next deliverable. Operating solutions are outside the scope of this book.

Some potential failures have been identified as the result of an initiating event somewhere in the plant's design, manufacture, installation, startup, or commercial phases. These can be controlled with quality assurance and control processes. This solution is actually part of the previous three.

Figure 8-2 shows that the final solutions may lead to iteration with the task of defining an ideal set of availability-centered practices (AB1.6, Chapter 5). Attention must be given to practices that were not previously recognized to be of much importance. Thus, this task provides verifying input.

AB9. Maintenance Logic-Tree Analysis

Definition and Purpose

The final task of FMECA identified four possible organizational solutions for each failure. One was for management by plant maintenance operations. The others were design, operations, and quality assurance practices. This deliverable is relevant when maintenance operations is the chosen solution.

There are four maintenance strategies to consider and select from for each failure. The choice will determine how effectively each failure will be managed. The plant's life-cycle cost structure will also be affected.

The maintenance logic-tree analysis is the best tool for determining the appropriate strategy.

Choosing between Maintenance Strategies

Four fundamental choices. The logic-tree analysis determines when maintenance is to occur. The four possibilities include:

- At fixed intervals (preventive maintenance).
- At a predefined condition or level of performance (predictive maintenance).
- During a failure-finding activity (inspection, servicing, and testing).
- In response to a failure event.

Preventive maintenance. Apparently, preventive maintenance is too often arbitrary rather than a choice derived by maintenance logic-tree analysis. One problem is that it can reintroduce "infant mortality" failures. This is especially true when complete overhauls are scheduled on the basis of time.

A task force was formed to investigate the effectiveness of preventive maintenance in aviation [Matteson, Donald, and Smith 1984]. One discovery was that time-based maintenance was only effective for a relatively small percentage of failures.

The study led to the following conclusions:

- In many cases, condition monitoring is much more effective than preventive maintenance.
- The ability to develop failure probability determines the effectiveness of preventive maintenance. Therefore, the gathering and analysis of performance data is a key to increasingly effective preventive maintenance.

Condition monitoring and surveillance.   Condition monitoring and surveillance is warranted when it is possible to manage equipment reliability by monitoring for problems. The ability to detect a trend of deterioration allows for planning rather than reacting to a sudden failure.

As a result, reliability may be extended for some equipment. This comes from assessing its current condition. A confident decision can then be made to continue use or to shut it down. Furthermore, the time to repair may be reduced by the opportunity to plan the necessary work.

Maintenance in response to a failure event.   The final choice is to recognize failures as they occur. In other words, do not attempt to prevent or detect them in advance, but run the equipment until failure.

The importance of identifying these cases may be indirect. They are opportunities with respect to planning and costs required for plant maintenance operations. In a few words, this choice also often identifies what is not significant. Some possible effects of this discovery on later detailed design of maintenance operations are as follows:

- Maintenance requirements can be less rigorously treated by maintenance task analysis (AD1).
- Procedures provided by manufacturers may be more readily accepted. Thus, they do not require the methodical development and management (AD2).
- The analyses of resources for these failures can be included as part of buffer levels in the calculation of resource provisioning instead of being meticulously detailed in the calculation of plant resources (AD4).

This is a helpful discovery since the alternatives to these possibilities have their costs, which are the costs incurred for the initial and periodic analyses and the management of their results.

Objective and Benefits of Maintenance Strategy Analysis

Objective.   The objective of the logic-tree analysis is to determine the most cost-effective set of maintenance strategies for the plant's failures. This is fundamental to achieving the life-cycle cost valley in Figure 1-10 produced by the following:

- Better alignment and application of maintenance resources.
- More appropriate ratio of preventive maintenance versus the other maintenance strategies.
- More appropriate use of testing, evaluation, and surveillance based maintenance.
- Breaking away from manufacturers' recommendations for preventive maintenance and other costly approaches. The experience of reliability-centered maintenance (RCM) practitioners has been that the source is a poor basis for decisions.

Historical results and findings.   RCM includes logic-tree analysis preceded by data development, plant logic diagrams, and FMECA. An airline's study found that its unit cost for aircraft maintenance remained constant for a 16-year period [Matteson, et al., 1983]. Safety performance steadily improved as well. The size and complexity of its aircraft also increased during that time.

Practitioners of RCM find that the alignment of the four maintenance strategies is typically inappropriate. Consequently, this suggests that the affected maintenance operations are substantially less effective and more costly than they need to be.

The Logic-Tree Structure

A logic tree should be developed for the analysis of each plant or facility. Its terminology and logic should be consistent with the terminology and focal concerns of the parent organization. However, there is a general pattern. It is shown in Figure 8-3 and discussed in the following sections:

Basic logic.   The basic branches of the logic tree are as follows:

- The first branch in the logic is whether the failure can be detected by testing and evaluation or by surveillance, and if it can be remedied prior to failure.
- If the failure cannot be detected and remedied, the logic may offer three possibilities. One is to use preventive maintenance to search for hidden failures. Another is to redesign the element or plant configuration. This is especially appropriate when a search for hidden failures is not possible. In these cases, there is no failure to detect until there is an observable failure.

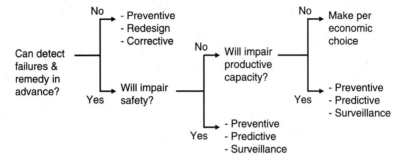

Note: Additional logic is applied at each branch to select the best choice from candidate alternate strategies at the branch.

Fig. 8–3 General maintenance logic-tree concept as an example

The third possibility is to respond with corrective maintenance as a failure occurs. This is possible when the failure is not ranked as highly significant.

- If the failure can be detected and remedied, analysis moves to the next branch. This branch is whether the failure results in a loss of function that will impair safety.
- If the failure *will* impair safety, the logic may offer three possibilities. They are preventive maintenance, predictive maintenance, and surveillance.
- If the answer to the safety branch is "no", the logic continues to the next branch. The focus is whether there are consequences for productive capability. If so, the logic leads to choices between preventive maintenance, predictive maintenance, and surveillance. If not, the logic leads to strictly economic choices.

The latter draws upon deliverables and their tasks to develop life-cycle costs. The life-cycle cost subsystem of the financial model is used in the deliberations.

In the "yes" case, the decision is also economic. However, the loss or the reduction of productive capability renders the outcome for more expensive methods to be generally automatic.

Sub-branches of maintenance choices.   Various branches offer different sets of maintenance strategies. These branches are extended to determine which is most effective. The logic of each will most likely be determined by the location in the master logic and the owner's circumstances.

It is expected that time-based maintenance (preventive maintenance) will often be the last choice. The remaining three will be chosen in accordance with some technical and economic rationale that is unique to the group of failures.

Tasks for Accomplishing the Deliverable of Performing Maintenance Logic-Tree Analysis

The tasks for performing maintenance logic-tree analysis are described in the following sections. They are flowcharted in Figure 8-4. The flowchart also shows the sequences and iterations with tasks in parallel and previous deliverables.

AB9.1. Formulate the decision logic.   A literature review will find many possible logic trees. This suggests that logic is ultimately unique to each application. Accordingly, this task is to formulate a logic tree for the subject facility.

AB9.2. Apply the logic to determine maintenance strategies.   This task is to apply the logic in team sessions. FMECA will have determined the failures to be analyzed. Figure 8-4 shows that the task also draws upon the analyzed economic data (AB3.5, Chapter 6) and the life cycle cost subsystem of the financial model (AB7.5, Chapter 7).

Figure 8-4 shows that there is potential for iteration from the later deliverable of assessing plant layout for maintainability (AB11.5, Chapter 7). It may be found that a selected maintenance strategy is not efficient or effective.

Figure 8-4 also shows that this task provides input to the determination of instrumentation design. Instrumentation is a traditional design deliverable (TB3, Chapter 10). However, maintenance logic-tree analysis will reveal instrumentation requirements for managing availability performance.

AB9.3. Establish cases of alternative solutions.   It is conceivable that some failures will have other feasible solutions. A final decision must be made with respect to profit-effective availability performance. This measurement is not in the domain of logic-tree analysis. The deliverable for availability analysis and optimization (AB10) serves that role.

Accordingly, this task identifies cases where there are alternative strategies. These are not just choices between the four fundamental maintenance strategies, they may also be choices within a single strategy.

AB9.4. Validate and adjust allocated availability parameters.   The completed deliverable provides new information for the design of availability performance. Thus, the final task is to review the previously allocated reli-

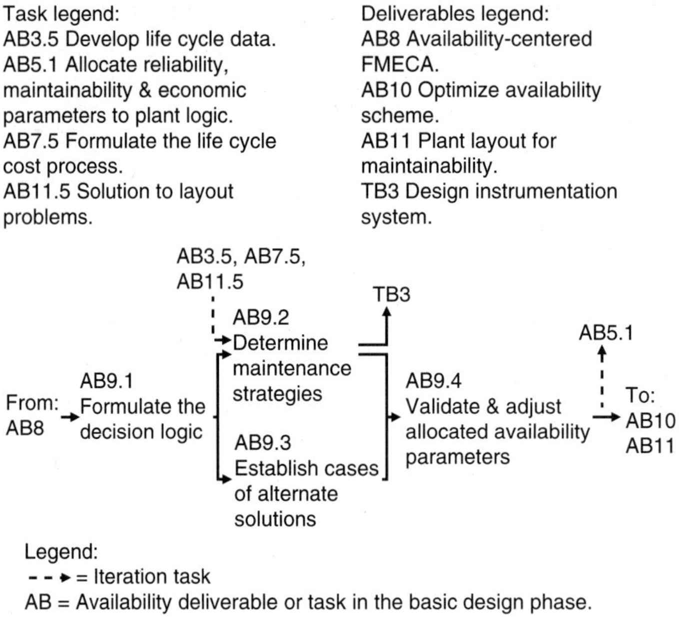

Task legend:
AB3.5 Develop life cycle data.
AB5.1 Allocate reliability,
maintainability & economic
parameters to plant logic.
AB7.5 Formulate the life cycle
cost process.
AB11.5 Solution to layout
problems.

Deliverables legend:
AB8 Availability-centered
FMECA.
AB10 Optimize availability
scheme.
AB11 Plant layout for
maintainability.
TB3 Design instrumentation
system.

Legend:
- - ► = Iteration task
AB = Availability deliverable or task in the basic design phase.

Fig. 8–4 Tasks for performing maintenance logic-tree analysis

ability, maintainability, and economic parameters (AB5.1, Chapter 7) to verify their validity or need for adjustment.

Bibliography

Blanchard, Benjamin S. *Logistics Engineering and Management*. 4th ed. Prentice Hall. Englewood Cliffs, N.J. 1991.

Bloch, Hienz P. and Geitner, Fred K. *An Introduction to Machinery Reliability Assessment*. Van Nostrand Reinhold. New York. 1990.

Electric Power Research Institute. Demonstration of Reliability Centered Maintenance. Volume 1: Project Description. Palo Alto, Calif. EPRI NP-6152. January 1989.

Jakuba, Stan. FMEA—A Reliability Tool That Works. Proceedings from the First International Conference on Improving Reliability in Petroleum Refineries and Chemical and Natural Gas Plants. Hydrocarbon Processing. Houston. November, 1993.

Matteson, McDonald and Smith. Commercial Aviation Experience of Value to the Nuclear Industry. Electric Power Research Institute. NP-3364. Palo Alto, Calif. 1984.

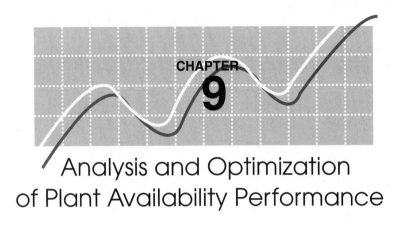

CHAPTER
9

Analysis and Optimization of Plant Availability Performance

Introduction

Availability engineering in the basic design phase has so far accomplished the following in the design of a performance scheme:

- Assessed reliability and maintainability of plant items, components, and assemblies.
- Allocated reliability, maintainability, and economic parameters to plant logic diagrams.
- Determined the cause, effect, and criticality of identified failures in the plant.
- Identified the most cost-effective maintenance strategy for each significant failure.

However, these are still preliminary conclusions. They have not been tested, evaluated, and optimized as a system of availability performance. That is now the objective.

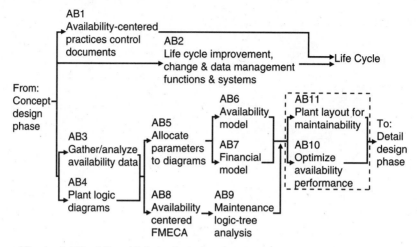

AB = Availability deliverable in basic design phase
FMECA = Failure modes, effects & criticality analysis

Fig. 9–1 Basic design phase availability engineering deliverables for the analysis and optimization of plant availability performance.

This is obviously a complex and challenging requirement. Figure 9-1 shows that optimization involves two deliverables during basic design. They are as follows:

- The testing, evaluation, and optimization of the plant availability scheme (AB10).
- Evaluation of the plant layout for maintainability (AB11).

AB10. Evaluate and Optimize Plant Availability Performance

Definition and Purpose

This deliverable applies the availability and financial models for determining an optimum availability scheme. Accordingly, it first defines the criteria and associated measures against which to determine an optimal scheme.

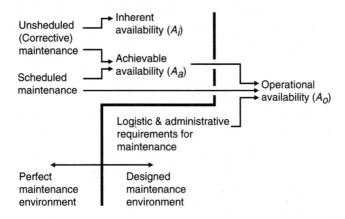

Fig. 9–2 Relationship of three types of availability.

Optimization Across the Three Types of Availability

Three types of availability. Chapter 1 introduced three types of availability. Their relationship is shown in Figure 9-2. To review, they are as follows:

- *Inherent availability* (A_i) is that which is expected when reflecting only unscheduled maintenance. Such maintenance is not included in a planned shutdown period. These periods are defined as those established by annual and longer-term planning.

  Inherent availability excludes associated logistic and administrative requirements. Thus, maintainability is computed only with active repair time.

- *Achievable availability* (A_a) is that which is expected when scheduled and unscheduled maintenance are combined in the calculation of availability. Associated logistic and administrative time is still excluded. Thus, inherent and achieved availability assume a perfect support environment.

- *Operational availability* (A_o) includes the logistic and administrative time in maintainability. Consequently, it incorporates decisions for plant maintenance operations in the calculation of availability. Therefore, operational availability is the bottom line of performance.

Optimization of achievable availability (A_a). Profitable plant availability is the result of the optimization of all three types. But what is optimization? It begins with achievable availability (A_a).

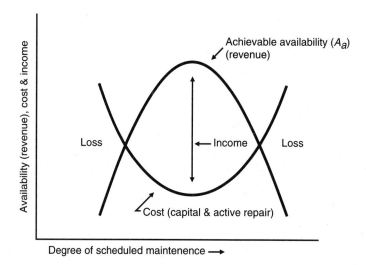

Fig. 9–3 Optimization of achieved availability and its costs.

First, it is necessary to more carefully define scheduled and unscheduled maintenance. *Scheduled maintenance* occurs in previously planned shutdown periods. It has been called time-based maintenance and preventive maintenance.

Unscheduled maintenance occurs when needed. This is done both immediately or in a preorganized event. Such maintenance may be assigned to a soon-to-arrive previously planned shutdown. Therefore, maintenance triggered by condition monitoring, surveillance, and failure falls in the unscheduled category.

The availability model can easily distinguish the plant elements subject to scheduled maintenance. Items subject to the strategy are shown as having an availability of 100 percent. The result is a model of availability performance between planned shutdowns (A_i). Therefore, achievable availability (A_a) is the product of two components. One is the percentage of time in a full cycle that the plant is to be in business. The other is its inherent availability (A_i) during the time that it is to operate. This is shown in Equation 1-2.

Figure 9-3 depicts achievable availability as a curve. It is the result of the following factors:

- Plant hard design determines the location and shape of the curve. Therefore, it establishes possible achievable availability.
- The maintenance strategies selected by maintenance logic-tree analysis (AB9, Chapter 8) determine the plant's location on the curve. Therefore, they establish actual achieved availability.

- The right extreme of the availability curve generally reflects 100 percent scheduled maintenance. It is a strategy of no surprises because all maintenance is performed in a scheduled shutdown period. Of course, it is not a real possibility. It allows only a description of the optimization problem.

  Availability is well below suboptimal. The plant's elements are not operated close or to the wearout point. An analogy is a car in a grueling and long race. A high percentage of cars do not mechanically survive the race. However, a car can be made to almost always survive the race by scheduled maintenance. However, each year the winner will have long ago taken the checkered flag.

- As scheduled maintenance is traded off against unscheduled maintenance, availability moves up the curve to the left. The movement is at first generally linear. This is because the elements are still not operated close or to the wearout point. Thus decreasing time for planned shutdown is rewarded one-for-one with availability performance.

- At some point along the curve, the result of the progressive trade-off is increasing at a decreasing rate. The system is now being increasingly challenged. More of its elements are being operated close or to the wearout point. The net result is still positive until a peak is reached.

- Left of the peak, achieved availability begins to fall. Additional trade-offs are now poor strategies.

  All things being equal, equipment provides maximum time-between-failure when operated to the wearout point. The curve falls because this ideal has its limits. Some failures will have undesirable results at the subsystem and system levels. For example, the permitted failure may cause others. These would not otherwise have occurred until later and possibly not at all.

The cost curve is equally important to this discussion. It reflects capital investment and active repair costs of achievable availability (A_a). At the left extreme, less plant investment is required. This is simply the result of greater service from each element in a period of time. Thus, a lessor system can still achieve management's specified performance.

Meanwhile, repair time and, therefore, maintenance cost is reduced for these items. This is because equipment requires less attention in a period of time. It is expected that overall cost is high rather than low at the left extreme. This is because the associated strategy also has many unattractive side-effects. As mentioned, some failures will create other very significant failures. Meanwhile, some selected equipment may be more costly to counter or protect against these side-effects.

The right side of the valley is skewed towards scheduled maintenance. This is also an extremely expensive strategy. Thus, the cost curve will rise and become the right side of a cost valley. The curve, therefore, develops a valley as the maintenance strategy progresses from one extreme to the other. Its shape and location are also determined by hard design.

The availability and cost curves are generalized by deduction. However, the relative positions of the extremes cannot be generalized here because they reflect the conditions of individual plant.

The availability and cost curves have another message. Availability is a proxy of revenues. Thus, at some point toward either extreme, the cost of availability performance will exceed the income it allows. Without availability engineering and management, operating beyond those intersections can occur without management's awareness. This is because normal accounting practices and other measures of maintenance operations cannot easily reveal that case.

Optimization of achievable (A_a) and operational (A_o) availability. The difference between achievable and operational availability is the inclusion of maintenance support. Achievable availability assumes that resources are always available and that there is no administrative delay in their application.

Therefore, maximum operational availability will theoretically "go to" achievable availability. In reality, there is a natural gap. Any endeavor has an upper limit of obtainable perfection. That natural limit is temporarily ignored for this discussion.

This relationship is shown in Figure 9-4. The lower curve is operational availability, the bottom line of performance. Its shape and location is a function of maintenance operation resources and organizational effectiveness.

Operational availability cannot rise above the upper curve. This is shown in Figure 9-5 for lateral points along the curves. Resources and organizational effectiveness can be increased to a point represented by the upper curve of Figure 9-4. Beyond that point, additional resources and organizational effectiveness have no effect.

Only then can availability be increased with a new set of maintenance strategies. That is if plant availability is to one side of maximum possible achieved availability. At the maximum, it can only be increased with new capital investment.

This is an important point. Without availability engineering and management, it is easy to unknowingly spend beyond that point. This may occur when plant performance falls short of management's desired productive capacity. Management tries to gain it with increased stress on maintenance support. However, the operational availability curve has already been unknowingly forced against the achievable availability curve. The result is

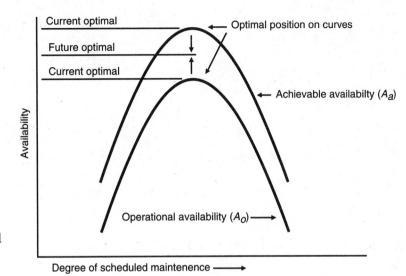

Fig. 9–4 Relative position and strategy for achieved and operational availability.

money down the proverbial "rat hole." Spending is in the loss zone to the right of the intersection of Figure 9-5.

Optimization choices. It is now possible to define the dimensions or nature of optimization. They are as follows:

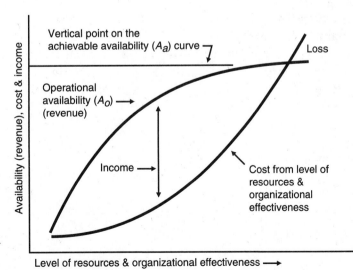

Fig. 9–5 Maximum possible operational availability for any point on the achieved availability curve.

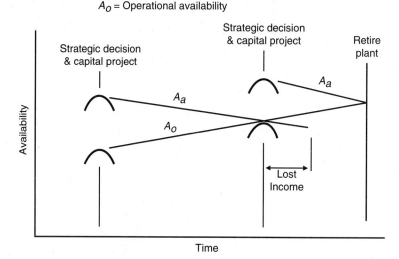

Fig. 9–6 Life-cycle trends in the positions of achievable and operational availability

- The location and shape of the achievable availability (A_a) curve.
- The choice of maintenance strategies for positioning plant performance at the peak of the curve.
- The level of operational availability (A_o) with respect to achievable availability.

The second dimension is an automatic goal. The others are a function of strategic issues. They will result in choices between the two extremes.

One possibility is capital versus working asset approaches to specified availability performance. The capital-driven approach designs the plant to have a high achievable availability (A_a) curve. It is then operated on a much lower operational curve. The working asset approach will have a lower capital investment. The plant is designed for the two curves to generally be much closer. In turn, the plant will be subjected to more frequent capital projects.

Optimization is not just a matter of performance at a point in time. It occurs over a period of time. In fact, one aspect of optimization is to establish that period and how the curves will be positioned and trended across it.

This process is demonstrated in Figure 9-6. Assume that management has elected to invest in capacity (A_a) beyond its immediate availability requirements. It then designs its maintenance operations to operate at the peak of the operational availability (A_o) curve.

Over time, the scheme for maintenance operation resources and effectiveness are adjusted along with the demand on the plant for productive capacity. This drives the operational availability curve upward. Meanwhile, the changing related operating conditions drive the achievable availability curve downward. Ultimately, the curves converge to a natural limit. From there, availability can only be increased by new capital investment.

Another example is the production facility of a large oil or gas field. Field production may increase for a significant number of years before beginning to decline. Thus, the achievable and operational availability curves may be designed initially to converge at peak production. Operational availability is adjusted along with required performance before and after that point.

These scenarios not only have many variations, but are dynamic. Thus, an optimal availability scheme is initially determined. It is then continuously evaluated and adjusted over the plant's lifetime.

Evaluation and Optimization of the Plant's Hard Design Issues

Evaluation and optimization across the segments of availability incorporate three dimensions. These are plant design, financial strategies, and the serving of market expectations.

The possible issues for plant hard design in the process of optimization are as follows:

- The first requirement is to determine whether the plant can actually achieve the specified achievable availability.
- Subsystems, equipment, assemblies and components are then ranked for their relative significance to availability performance. This will determine where the design team's attention and energies will have the greatest value.
- Experienced engineers often design redundancies into the plant. These safeguards of performance might be justified, or they could also be an expensive over-design. Therefore, availability evaluation attempts to confirm that all redundancies are warranted.

  There is a related possibility. The redundancy is necessary, but the strategy is inappropriate. For example, redundancy may be arbitrarily set at 100 percent instead of a derived determination.

- Some allocated reliability and maintainability parameters may turn out to be physically and economically unrealistic. Such cases may appear as the plant design progresses. Other may be discovered during the procurement process. The availability modeling process is used to formulate solutions and an adjusted optimum.

- Availability assessment is critical to the decision processes to accept proposed design changes. The design team must be able to analyze complex technical and financial balances in a timely fashion. Management must decide whether to delay a project completion date, make an additional investment, or defer and possibly forego the opportunity associated with the proposed change.

Evaluation and Optimization for Financial Strategies

The previous section identified plant hard design issues for evaluation and optimization. Another identified dimension is financial strategy. It spans inherent, achievable, and operational availability. These include strategies to shift life-cycle costs from investment to functioning, maximization of income, maximum productivity of assets, etc. Some strategic cases for financial evaluation and optimization are:

- A plant may be designed and built when time-to-completion is strategically crucial. An example is the opportunity to begin commercial production in a strong business cycle. Another condition that presents similar constraints is when capital funds are limited.

  Availability analysis must determine which availability performance opportunities the owner can afford to permanently lose. The performance value of "now or never" opportunities are weighed against the strategic implications of a later project completion.

- There are many choices of capital versus working assets balances that will achieve the same availability performance. These range from capital (total installed cost) solutions to working asset (maintenance operations) solutions.

  Analysis and optimization treats this case from two directions. One is to find the optimal balance between capital and working assets (scheme for no waste of assets). Alternately, it may be necessary to determine the penalty of an availability scheme to one side of the financial optimum. This is the case when there are strategic and financial constraints to be served.

Optimization for Market Expectations

Design for long-term market expectations is also a dimension of evaluation and optimization. This is because plants must change over time in response to market trends.

The possibilities are actually integrated with the other two dimensions. There are two basic strategic choices in the overall optimization. They are as follows:

- Capital investment in availability as demand increases.
- Investment in availability capacity up-front to meet expected long-term market trends.

In both cases, operational availability performance is increased along with demand trends. Time-to-maintain is progressively decreased. The associated still-flexible factors of reliability and maintainability are adjusted through maintenance operations. The difference is that the time between the capital investments required to match the market will vary.

Basic Methods for Testing and Optimization

There are as many approaches to testing, evaluation, and optimization as the imagination will allow. However, there are universal requirements or principles that determine their legitimacy. A legitimate process should include the following:

- A determination of the relative sensitivity of availability performance to plant elements.
- The formulation of a set of candidate schemes.
- Determination and relative ranking of the performance criteria.
- Identification of the performance measures for each criterion.
- Measures for the criteria set must be formulated as a single formula. The formula is then used to value candidate availability schemes in search of the best one.
- Identification and further assessment of the best candidate for the consequences of major changes.

These concepts are included in the tasks described in the following section for evaluating and optimizing the plant availability scheme.

Tasks for Accomplishing the Deliverable of Evaluating and Optimizing the Availability Scheme

The tasks of analyzing and optimizing the availability scheme are described in the following sections. They are flowcharted in Figure 9-7. The flowchart also shows the sequences and iterations with tasks in parallel and previous deliverables.

AB10.1. Establish criteria to evaluate the availability scheme. Criteria must be established as a basis against which the availability scheme can be evaluated and optimized. Therefore, their selection is a critical process. This is because an important but omitted criterion may be lost in the choice of the best scheme.

The primary source of criteria is the conceptual design phase deliverables. The careful inspection of them during the performance of this task will reveal important criteria.

The revealed criteria are operational and financial. Operational criteria may stress reliability versus maintainability, relative staff levels, stress on hard design versus maintenance operations, basic position among the three types of availability, etc. Financial criteria may stress income maximization, maximum asset utilization, optimization between income and asset utilization, availability performance from capital versus working asset strategies, etc.

Task legend:
AB5.1 Allocate reliability, maintainability, and economic parameters to plant logic.
AB6.6/7 Execute plan & use availability model.
AB7.8/9 Execute plan & use financial model.

Deliverable legend:
AB6 Develop plant availability model
AB7 Develop plant financial model.
AB9 Maintenance logic-tree analysis
TB1 Refine the conceptual design.
TB4 Performance specifications, size & select equipment.
TD3 Procurement of engineered equipment & items.

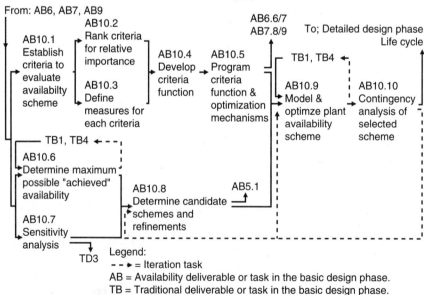

Fig. 9–7 Tasks for evaluating and optimizing the availability scheme (AB10 of Figure 9-1).

AB10.2. Rank criteria for relative importance. The identified criteria are not equal in importance. This task is to rank them. Ranking is relative rather than cardinal. The steps are as follows:

1. Rate each criterion for its importance in an optimal scheme. Each is rated on a scale of 0 to 10.
2. Establish relative rank by weighted average techniques.

AB10.3. Define the measures of each criteria. Criteria are not always directly measurable. Thus, measures must be determined for each. Unlike criteria, they can be extracted directly from the performance models and databases. For example, a measure of high overall reliability as a criterion is maintenance hours per operating hour. A measure for the criterion to optimize across income and asset utilization is return on investment.

A criterion may have multiple measures. The same measures often serve multiple criterion.

AB10.4. Develop the criteria function. Initial design and then changes throughout the plant's life will always have multiple candidate solutions. Thus, there must be a means to choose between them. The method must also be able to value them relatively.

The criteria function serves this purpose. It synthesizes the criteria and their set of measures. The result is a single equation that will compute a value for each candidate. It is expected that the highest (or lowest) value will be generated by the best candidate.

There are many approaches to formulating a criteria function. This book will not attempt the subject. Ostrofsky [1977] has developed a rigorous mathematical approach for large, complex systems.

AB10.5. Program the criteria function and optimization mechanisms in the models. At this point, the availability and financial models should already be in the process of development. They should at least be at the stage necessary to serve ongoing project needs. At this stage of development, there is a capability for optimization. This requires the following steps.

*   Program the criteria function in the pertinent models and their subsystems.
*   Determine how candidate availability schemes will be presented and drawn into the models.
*   Determine the models' optimization processes and the presentation of results.

- Extend existing plans and activities to both models to include the optimization processes. These tasks are part of the deliverables for developing the availability (AB6) and financial (AB7) models. One is to set up both models for timely plant design (AB6.6 and AB7.8). The other is to execute the plans to fully develop the final models (AB6.7 and AB7.9). Accordingly, Figure 9-7 shows the results of this task flowing to the models.

AB10.6. Determine the plant's maximum possible achievable availability (A_a). In the previous section it was mentioned that an objective of hard design is to create the plant's capacity to reach a specified achievable availability (A_a). If the hard design has produced a plant that cannot reach that level, it cannot be reached even in a "perfect" maintenance environment of infinite human and material resources and organizational effectiveness.

This task is to determine if the design is acceptable. This means that the peak of the A_a curve in Figure 9-3 should be at or above the performance specified in the conceptual design phase. This determination was first made when preliminary parameters were allocated to the plant diagrams (AB5.4, Chapter 6).

Accordingly, Figure 9-5 shows that the findings of the task will be iterative with traditional design. The most affected deliverables are:

- The final conceptual design (TB1) and subsequent plant engineering diagrams.
- The performance specifications, size, and selection of equipment (TB4).

AB10.7. Sensitivity analysis. A principle of optimization is that all feasible candidate must be identified for evaluation. Availability candidates exist if there are possible reliability and maintainability variations for any plant element.

The collection of previous deliverables will have revealed candidates schemes along the way, but additional attractive candidates may still exist. The models are used to determine where to search for them in the availability scheme.

This is done with sensitivity analysis defined as the rate of change of plant availability performance in response to changing a plant item. Of greatest interest are cases that fall at the extremes of high and low sensitivity.

The search for points of opportunity is easily accomplished. The model is run with each plant item, in turn, set to some extreme. Some possibilities are:

- Availability is set at 100 percent.
- Investment costs are set to zero as availability is set at an arbitrary low figure.

- Functioning costs are set to zero as availability is set to an arbitrary low figure.

The last two tests relate cost and availability sensitivity. It is not important to have an exact figure. The goal is to create a sense of relative sensitivity.

Figure 9-7 shows that the findings will flow to traditional procurement processes (TD3, Chapter 14) in the detailed design phase. Sensitivity may indicate a basis of choices between competing suppliers. The choice may be made according to a supplier's experience with components and assemblies.

AB10.8. Determine candidate schemes and refinements. The findings of the sensitivity analysis are applied in two ways prior to modeling for optimization. They are as follows:

- Evaluate and determine reliability, maintainability, and economic variations for items associated with high sensitivity. The objective is to develop a set of alternatives for each. These generate new candidate availability schemes.
- Evaluate the items found to have low sensitivity. The objective is to determine if their treatment in the availability scheme can somehow be reduced. The result is to accept a lesser reliability and maintainability performance with respect to plant items. This is because plant availability performance is not sensitive to this reduction. Meanwhile, life-cycle cost is reduced. The same dollar spent on the higher sensitivity items has a much greater payoff in availability performance.

The adjustments for low sensitivity are made before optimization. They are refinements rather than candidates.

The candidates and refinements will require an iteration with the task to allocate reliability, maintainability, and economic parameters to plant logic diagrams (AB5.1, Chapter 6). The candidates become alternate sets in the system of availability performance parameters.

AB10.9. Model and optimize the plant availability scheme. The final requirement is to model the plant. The objective is to search out the best candidate scheme. It is revealed by the criteria function installed in the availability and financial models.

Figure 9-5 also shows that this task is primarily iterative with two traditional deliverables (TB1 and TB4). These deliverables are the same as those in the above analysis loop of maximum possible achievable availability (A_a).

However, there is a another purpose. The earlier iterations were to confirm that the owner's specifications can in fact be met. The iterations of this task and the traditional deliverables are to find the most profitable scheme.

AB10.10. Contingency analysis of selected scheme. Contingency analysis should now be applied to the chosen scheme. The purpose is to evaluate the consequences of change in operational requirements and conditions. These changes may have many sources. Examples are long-term demand trends, business cycles, new technology, new and evolving regulations, developments in the owner's larger production system, etc. The objective is to evaluate the validity of the chosen scheme in the context of possible scenarios.

This task is iterative within the deliverable. It may lead to new candidate schemes or reshape the choice of the optimal scheme.

AB11. Assess Plant Layout for Maintainability

Definition and Purpose

Up to this point, the deliverables have assumed that the plant layout for maintainability is perfect. Therefore, the purpose of this deliverable is to evaluate plant layout for maintainability. The work products of previous deliverables provide a focus for the assessment.

Determinates of Maintainability in Plant Layout

One determinate of the plant's characteristic maintainability is its layout. The factors that affect maintainability are space, location, arrangement, accessibility, equipment handling, working environment, and color-coding and labeling. This will not be an in-depth discussion of human factors and plant layout. However, the pertinent factors are defined as follows:

- *Space* includes pathways, laydown areas, pull-space, working areas, and support facilities.
- *Location* is the distance and barriers between the work point and maintenance facilities, equipment, spare and repair parts, *etc.* The design for location includes decisions for central versus local placement, security boundaries, etc.
- *Arrangement* is the placement of equipment, components, piping, and other elements for maintainability.

- *Accessibility* is the provision for reaching the work location with ladders, hatches, platforms, etc.
- *Handling* is the provision for lifting and moving equipment. This includes planning for the movement of large components from installed or yard locations to positions within workshops. This is especially crucial for major equipment items or its pieces.
- The *working environment* is the concern for the physical aspects of the plant with regard to maintenance efficiency as a function of stress and fatigue from extreme temperature, humidity, and the use of protective clothing. The issues include temperature and humidity, noise levels, and lighting.
- *Color coding* and *labeling* guide maintenance operations, planning, and field activities.

Special Aspects of Maintainability Design

Some interesting aspects of the plant layout are touched upon in the following sections. These are support facilities, surveillability and accessibility, movement of large items, maintenance environment, and color coding and labeling.

Location and footage of support facilities. Location and footage of support facilities is part of the plant layout design. The nature and timing of the capital project does not allow a neat sequential approach to its design. This is because the following data and information is limited until the detailed design phase:

- Location, volume, and frequency of tasks requiring support facilities. This is provided by maintenance tasks analysis (AD1).
- Human and material resource levels to be provisioned for plant functioning (AD4).
- Detail and location of the management and administrative functions that will reside in the plant (AD5).

Expectations for support facilities were first established during the conceptual design phase. This was part of identifying the major elements of maintenance operations (AC3). Top-level utilization measures were also established.

Later, these may be "sanity-checked" by consensus. However, designers must be careful when using consensus and precedent as the primary design tool. A more rigorously developed preliminary design process may be taken. A possible approach is offered in the later described tasks for assessing plant layout for maintainability.

Surveillability and accessibility. Bloch and Geitner [1990] stress the importance of surveillability. Plant layout for accessibility and the associated capacity for surveillance, the ability to monitor deterioration, and access to detection devices are all extremely important.

This contention is reinforced by the principles of maintenance logic-tree analysis. The issue of the first branch was the ability to inspect a condition or detect a fault. Thus, poor surveillability and accessibility in plant layout may restrict the options for maintenance strategies.

Movement of large items. Evaluation for the movement of large items is a special problem. "War stories" suggest that this is not usually well planned. One reason is a major feasibility criteria for any proposed plant improvement. They must often be achievable within the time allowed for planned maintenance downtime.

A study of a living availability improvement program described such a problem [Unkle and Hall, 1989]. The program's purpose was to improve availability in an actual plant.

Criteria were established to screen candidate capital investment projects. One criterion was that it be accomplished within the time allowed for planned plant and subsystem shut-downs. Candidates were also screened for economic value.

A significant level of funds was to be budgeted for each year. It quickly became difficult to expend the planned annual budget. This was because the time criterion eliminated so many candidates. However, there was always plenty of economically attractive possibilities.

The point is that plant layout must allow for the movement of large equipment or its parts. Accordingly, life cycle constructability engineering should be included in plant design. The resulting layout may allow more improvements than would otherwise be possible.

This will require equipment removal worksheets that provide a step-by-step accounting of access, movement, and removal time. It may also require equipment removal isometrics that depict a necessary removal path and volume.

The traditional view of constructability engineering has been limited. The goal has been to determine how to reduce plant investment. The resulting design is easier and, therefore, less costly to construct.

The working environment. Maintenance and safety is affected by the plant's working environment. Troublesome environmental conditions affect maintenance performance by impairing the senses, mental activity, and the use of hands and fingers.

The capacity of the environment to allow the worker to stay focused is important to safety. This is to protect people from equipment and equipment from people. Thus, the plant layout for the working environment should consider the following safety options:

- Can a design change eliminate exposure to a hazard?
- Are there options for removing people from the hazard? Possibilities include robotics and instrumentation.
- Can guardrails be placed in the plant if the previous two possibilities are exhausted?

These questions are important to availability design because they can affect reliability and maintainability. Reliability is reduced if a reduced quality of repair work increases the probability of equipment failure. Maintainability is affected if the time planned to complete a task must be increased.

Labeling and color-coding systems The labeling and color-coding systems are immensely important issues to plant layout for availability management. Their objectives are as follows:

- Reducing maintenance time from delays in search of equipment.
- Reducing the possibility of overlooking a stop during routine maintenance and servicing.
- Notifing personnel of special actions and procedures.
- Avoiding the consequences of operating or repairing the wrong plant item.
- To function as a subsystem in the collection of availability engineering and management data and information.

Tasks for Accomplishing the Deliverable for Assessing the Plant Layout for Maintainability

The tasks of analyzing plant layout for maintainability are described in the following sections. They are flowcharted in Figure 9-8. The flowchart also shows the sequences and iterations with tasks in parallel and previous deliverables.

The figure also shows that there is input to this deliverable from traditional design. The traditional work scope includes the general development of the plant layout.

The following availability tasks are integral to that development. Consequently, iteration with the traditional layout process is shown as input and output of this deliverable.

AB11.1. List all layout-critical points of the availability scheme. At this point, the layout-critical points of reliability and maintainability are known or evolving. As mentioned, they have been revealed by the other phase deliverables. This task is to develop a list of all such points.

AB11.2. Assess plant layout against layout-critical points. Assess the plant layout against the identified critical points. The result will be a focused, rather than a general, plant assessment with respect to availability performance.

Figure 9-6 shows that availability-centered practices documents (AB1.8) are inputs to layout assessment.

AB11.3. Assess location and space of maintenance support facilities. The location and layout of maintenance facilities are crucial to maintainability. It was mentioned earlier that this is a difficult requirement during the basic design phase. However, it is still important to estimate location and footage.

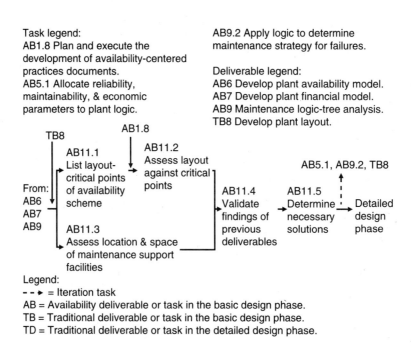

Fig. 9–8 Tasks for assessing plant layout for maintainability (AB11 of Figure 9-1).

A process for this purpose is as follows:

- Tabulate the possible maintenance tasks for each piece of equipment.
- Identify which tasks will significantly involve support facilities.
- Estimate the human and material resources for each task.
- Estimate the frequency of each task and its required time-to-complete. The allocated availability parameters (AB6) may be used.
- Determine support facilities in terms of location and estimated space requirements. The steps are as follows:

  1. Sort the tasks according to the anticipated facility.
  2. Use probability methods to synthesize frequency and time requirements.
  3. Use industrial engineering and architectural design methods to estimate space.
  4. Account for life-cycle stages and their needs. These include plant construction, startup, and producing life.

- Explore the implication of sharing new or existing shops with other plant units. The advantages include shared equipment, skills, and technical documents. The disadvantages are the subtle differences in necessary design, site layout, and scheduling. There is also a potential problem of maintaining necessary separation and identity of equipment and parts.

AB11.4. Extend the plant assessment to validate the findings of previous deliverables. The results are used to validate the findings of previous deliverables. Of specific interest are the allocated parameters of reliability and maintainability. Also of interest are the maintenance strategies selected by logic-tree analysis.

AB11.5. Determine and select necessary solutions. The preceding tasks will reveal inconsistencies between the availability scheme and evolving hard design. They are evaluated to determine a remedy. There are three possibilities. The following may be applied separately or as an integrated solution:

- Change the plant elements that are not the result of hard design. The reliability and maintainability parameters will change accordingly (AB5.1).
- Change the failure response strategies (AB9.2).
- Change the plant layout (TB8).

Bibliography

Bloch, Hienz P. and Geitner, Fred K. *An Introduction to Machinery Reliability Assessment*. New York, Van Nostrand Reinhold, 1990.

Buffa, Elwood S. and Sarin, Rakesh K. *Modern Production Operations Management*. 8th ed. New York, John Wiley & Sons, 1987.

Ostrofsky, Benjamin. *Design, Planning and Development Methodology*. Englewood Cliffs, N.J., Prentice Hall, 1977.

Pack, R.W., Seminara, J.L., Shewbridge, E.G., and Gonzalez, W.R. Human Engineering Design Guidelines for Maintainability. General Physics Corporation. Atlanta, Ga., 1985. EPRI NP-4350.

Sanders, Mark S. and McCormick, Ernest J. *Human Factors in Engineering and Design*. 7th ed. New York, McGraw Hill, 1993.

Unkle, C.R. and Hall, S.C. Demonstration of an Availability Optimization Method. For the Electric Power Research Institute by ARINC Research Corporation, February 1989. EPRI NP-1567.

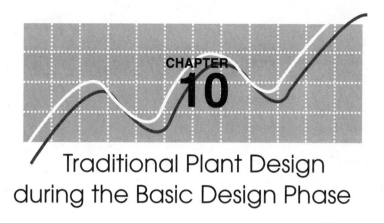

Traditional Plant Design during the Basic Design Phase

Introduction

It is helpful to describe the traditional plant design process because some readers may not be generally acquainted with the traditional capital project deliverables. This is important because availability design deliverables are bilaterally integrated with them. For the existing plant, this vantage helps reveal which aspects of its design must be evaluated for their affect on availability performance.

This phase in the life cycle has various names. They include basic design, preliminary design, Phase II design, etc. However, it is best defined by what preceded it and what it will produce.

The traditional conceptual design phase will have produced the following work products:

- A selected production process.
- The associated production configuration and equipment defined to an appropriate degree.
- Conceptual design results documented as flow diagrams, production process performance parameters, equipment configuration, and plant layout.
- Narrative documents that explain the production process and plant design issues and decisions.

- Design issues to be resolved or optimized during the basic design phase.

In turn, the traditional basic design phase will generally include the following:

- Studies to optimize subsystems in the production process and plant scheme.
- Detailed plant and equipment layouts.
- A scheme and layout for process monitoring and control.
- Utility distribution schemes and layouts.
- Equipment specifications.
- Safety analysis for identifying hazards in the design.
- Procurement activities for long lead-time equipment.
- Final diagrams, layouts, and major equipment specifications for owner approval.

Historically, these work products have given only brief attention to availability engineering and management issues. Treatment is generally limited to good practices, service levels in equipment selection, and recommendations for parts stocking. There is often some reference to procedures and training. However, this often lacks structure and is skewed toward process rather than maintenance operations.

The traditional deliverables of the basic design phase are flowcharted in Figure 10-1. The deliverables are coded as "TB" to denote a *traditional* deliverable during the *basic* design phase. Each deliverable is described below.

The flowchart also shows input, output, and iteration with the availability engineering deliverables. These items are coded as "AB" to denote an availability deliverable during the "basic" design phase.

It is not the intent of this chapter to rigorously describe and flowchart the tasks of the traditional deliverables as they were for availability deliverables.

TB1. Refine the Conceptual Design and Related Diagrams and Data

Traditional Scope

The conceptual design phase identified design aspects to be further explored and optimized. These are typically subsystem rather than system-level design requirements. Accordingly, this deliverable develops the final

Task legend:
AB1.8 Plan & execute decisions for availability-centered practices documents.
AB3.4/5 Develop reliability, maintainability and life-cycle cost data.
AB3.7 Define system for continual collection & processing of availability data.
AB4.1 Gather pertinent traditional design work products.
AB5.4 Validate the allocation of availability parameters (performance).
AB8.4 Determine the cause of each significant failure in the plant.
AB8.8 Determine organizational solution to significant failures.
AB9.2 Apply logic tree to determine maintenance strategies for each failure.
AB10.6 Determine plant's maximum possible achievable availability.
AB10.9 Model & optimize the plant availability scheme.

Deliverable legend:
AB11 Assess plant layout for maintainability.

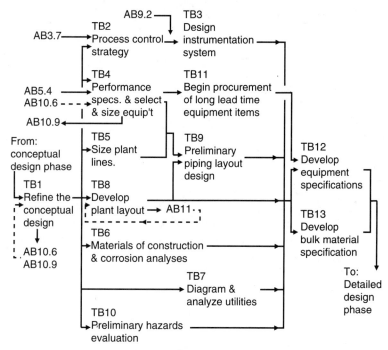

Note: Most traditional deliverables have input from availability
deliverables AB1.8, AB4.1, AB8.4 AB8.8 and AB3.4/5.

Fig. 10–1 Traditional deliverables for the basic design phase.

conceptual detail of the process and its production scheme.

Its possible activities are as follows:

- Reviewing the process design and associated mechanical flowsheets and site plans with the owner.
- Completing any special problems left open by the conceptual phase studies and design.
- Acquiring approval for the final results.

Pertinence of Availability Design

If availability engineering is part of a project, special problems usually left to the traditional disciplines would now involve the availability discipline. This may be developing or reviewing each solution for availability performance.

Recall that the conceptual phase translated top-level elements and measures of availability to the plant subsystems. They would now be adjusted in response to the refinements being explored by this deliverable. Without the discipline, they are only intuitive with respect to availability performance.

Figure 10-1 shows three primary relationships with availability tasks. They are as follows:

- It would provide plant details to the tasks that develop reliability, maintainability, and economic data (AB3.4 and AB3.5, Chapter 6).
- It would be involved in the confirmation that the evolving plant design can actually perform at a specified availability (AB10.7, Chapter 9). This may be an iterative process.
- It would be involved in iterations to optimize availability performance (AB10.9).

The last two relationships will reflect project progress. Consequently, the basic design-phase availability deliverables are done in loops. An analogy is project cost estimating. As detail increases, progressive estimates are made. The confidence limits are narrowed with each rendition.

As the plant design progresses, so does its detail. At some point during the project, plant design becomes fixed. It then becomes a "given" to the optimization cycle. Thus, the bilateral interaction between the traditional and availability engineering deliverables is a window of opportunity.

TB2. Determine and Develop the Process Control Strategy

Traditional Scope

The production process must be controlled to achieve its intended input/output ratios. Control is also necessary to achieve continuous periods of production at the planned equilibrium.

Levels of control. The design team's objective is to serve the following three related control needs:

- Control of process subsystems. This is the role of distributive control system (DCS) technology and associated advanced process-control software. Its purpose is to achieve and maintain a local optimization.
- Control of the total process. This is a global optimization of the process. Thus, it supervises the DCS controlled subsystems as a group. A relevant technology is Supervisory Control and Data Acquisition (SCADA).
- Control of the shutdown control process.

Goals of control. The overall control system is designed to serve the following objectives:

- Product quality in terms of meeting product specifications.
- Economic and productivity optimums and targets.
- Process operation parameters within boundaries that are related to safety and environmental hazards.
- Process operation within the limitations of equipment performance, reliability, etc.
- Production process operation within parameters its operators feel comfortable with.

Pertinence of Availability Design

If availability engineering is part of the plant design scope, the control system can be assessed with respect to the following:

- Assuring that equipment will be operated within boundaries for achieving management's specified availability. This would be defined by the task of optimizing availability performance (AB10.9, Chapter 9). This also suggests that process optimization would reflect operating conditions and the consequences for reliability and, in turn, availability.

- Data acquisition from SCADA would be planned to accumulate identified reliability data. The deliverable would receive input from the task of defining a data system that would continuously collect, access, and process data over the plant's lifetime (A 33.7, Chapter 6).

TB3. Design the Instrumentation System

Traditional Scope

The instrumentation system measures process performance with respect to flow, composition, temperature, pressure, level, etc. The measurements are determined by the control system design. They are the feedback from which the operator or control system will determine necessary control actions.

Pertinence of Availability Design

Issues of availability performance are conspicuously absent from traditional instrumentation design. A survey of power plant personnel [Seminars and Parsons, 1981] researched this issue. Plant instrumentation was found to be skewed toward process operation. The needs of maintenance were not usually considered.

However, the deliverables of availability design identify requirements that could be served by instrumentation. The most visible is the one for determining maintenance strategies for each significant failure (AB9, Chapter 8). One ultimate result is to determine how each failure is detected (AB9.2).

The points are revealed where condition-based monitoring and routine surveillance are the best means to identify the need for maintenance actions. Thus, availability engineering would methodically identify individual failures and how they are to be detected. This allows detection to be part of the overall instrumentation scheme. Otherwise, instrumentation for availability performance would be based only on intuition, experience, and history.

TB4. Develop Performance Specifications, and Select and Size Equipment

Traditional Scope

The plant equipment items are defined for performance and then sized by:

- Converting the parameters of the production process to the performance specifications against which equipment can be selected. Examples are rate, cycles, levels, temperature, pressure, etc.
- Establishing performance specifications for efficiency, service levels, operating conditions, etc.,
- Determining equipment sizes as a function of the performance specifications.

Whether the project design engineers actually select and size equipment or only develop performance specifications will vary. For some equipment, the project engineer selects or sizes the equipment. For others, the engineer only provides performance parameters. The equipment manufacturer makes selection and sizing decisions. These are based on their specialized expertise.

Pertinence of Availability Design

Sizing for service levels and operating conditions is central to availability performance. Who is to know what the full set of appropriate availability performance parameters should be? The discipline would supply such information.

Better yet, the information would depict the plant as a system of parameters. Without system-centered availability parameters there will be two unavoidable consequences. The service levels will either be excessive or insufficient with respect to the specified availability performance. Both result in reduced income and productivity of capital and working assets.

Figure 10-1 shows the relationship between the traditional deliverable and the availability tasks. They are as follows:

- The deliverable would provide plant detail to guide the analysis and development of reliability, maintainability, and economic data (AB3.4 and AB3.5, Chapter 6).
- Reliability, maintainability, and economic parameters will become available (AB5.4, Chapter 6) to the specification of the service level. Thus, they reflect both availability and process performance.
- Like the earlier deliverable for refining the conceptual design (TB1), this deliverable would be part of the iteration to first verify, then optimize, availability (AB10.6 and AB10.9, Chapter 9).

The degree of interaction is a function of plant design progress. Initially, there would be bilateral influence. With time, the design becomes a "given" to the continuing optimization. In other words, a window of opportunity is forever closed when availability engineering is not part of the original design phases.

TB5. Size Lines

Traditional Scope

For some plants, piping is a considerable part of the system. When this is so, plant piping is sized and evaluated for stress. The activities are generally as follows:

- A list of process lines and materials is developed.
- Line sizes as a function of flow ranges, pipe flow surface, throughput properties, and pressure drops are calculated.
- The consistency of calculated solutions with the equipment sizing results is confirmed.
- Which lines will require rigorous stress analysis is determined.

Pertinence of Availability Design

Availability engineering is pertinent to this traditional deliverable. The failure modes, effect, and criticality analysis and the analysis of availability related data and information contributes to this deliverable. They identify where stresses on equipment from pipe design can reduce reliability.

The traditional design process also seeks out such cases. It then designs them out of the plant. Therefore, FMECA and data analysis offers an independent search for stress problems.

TB6. Determine Materials of Construction and Corrosion Allowances

Traditional Scope

Like piping, materials and corrosion are a significant concern to some types of production plants. Materials engineering determines and specifies the appropriate materials for all points in the production process. It also specifies their corrosion allowances. Alternate strategies are determined when allowances are unacceptable. These include the use of alloys and injection of inhibitors in the process stream.

Pertinence of Availability Design

Availability engineering relates to materials design as follows:

- Availability-centered FMECA identifies when materials are especially relevant to the reliability of some equipment (AB8.4, Chapter 8).
- Materials may also be relevant to maintainability if special practices and care are necessary. FMECA would isolate these cases if they create an additional potential for failure.
- Life-cycle cost analysis reveals the impact of special materials on the economics of maintenance operations versus investment (AB3.5 Chapter 6).
- Preliminary modeling reveals when the effect on reliability or maintainability is significant to plant availability performance.

TB7. Diagram, Analyze, and Determine Utility Needs

Traditional Scope

There are three requirements for the design of the plant utilities scheme.

1. Develop flowsheets for a scheme to serve the plant with utilities. They identify the components for providing electrical, water, air, steam, inert gases, fuels, refrigerants and heating substances.
2. Undertake load and level studies. They determine the timing and fluctuations of demand for each utility.
3. Determine the requirements for tankage, lines and sizes, instrumentation and controls, equipment specifications, etc.

Pertinence of Availability Design

Utilities design can affect plant maintainability and, thus, availability performance. Accordingly, availability engineering will be involved in the utilities development design process. Its deliverables will identify maintenance tasks, location, and their needs. Otherwise, the process of determining utilities for maintenance events would follow design checklists rather than reviewing maintenance task needs.

Admittedly, in most cases the result may be the same. In other cases, it may be found that it is necessary to go beyond standard practices.

TB8. Develop the Plant Layout

Traditional Scope

The plant layout includes:

- Plot plans.
- Equipment layouts.
- Electrical and hazardous area classifications.
- Basic piping location and routing.

Layout design is an iterative process that spans the engineering design phases. Progress dictates that it begin in the conceptual design phase. During basic design, pertinent information is either incomplete or not yet final. This is because equipment dimensions will not be final until the advanced stages of procurement and detailed design.

Pertinence of Availability Design

Availability engineering will affect this deliverable through its attention to maintainability. The work products of availability design offer an opportunity for a focused review rather than general review. Accordingly, the contributions to layout design are as follows:

- Preliminary determinations of locations and sizes of maintenance support facilities (AB11.3, Chapter 9).
- Checklists applicable to plant layout design for maintainability were developed (AB1.8, Chapter 5).

TB9. Design Preliminary Piping Layout

The above production equipment, their sizes, and plant layout lead to preliminary piping layout design. The primary purpose is to validate the layout. The product will probably be limited to single line diagrams of the plant piping. Meticulous design is part of the later detailed design phase.

TB10. Conduct Preliminary Hazard Evaluation

Traditional Scope

Assessment and identification of hazards determine the fundamental risks that are associated with the evolving design. At this stage, they can only be preliminary. Analysis should be consistent with the current level of design detail.

Hazards must be identified for two reasons. First, so they can be considered in the subsequent design activities. Second, for the planning and implementation of production process safety management and control functions.

There are various techniques available for preliminary hazards analysis. Possible methods are checklists, preliminary hazards analysis, and "what-if" analysis. All of them would make use of evolving availability analyses and detail.

Fortunately, much attention is now being given to process safety management. However, neither this deliverable nor the one for rigorous hazards analysis in the detailed design phase is process safety management. They are only a part of a larger management concept. The main objective is to develop a holistic structure of functions, processes, and systems for controlling and managing plant safety. Preliminary and detailed evaluation of hazards are both within that domain.

Pertinence of Availability Design

The work scope, insight, and deliverables of an evolving availability design will aid initial hazards assessment. Its contributions would be as follows:

- The discipline would be in the process of gathering and analyzing reliability and maintainability data (AB3.4 and AB3.5 of Chapter 6). These are prerequisite to any quantitative hazards assessment.
- The web of failures and their consequences for plant, subsystem, and equipment availability is an evolving analysis (FMECA, Chapter 9). Hazard related failures are included.
- The plant availability concept will have determined the nature and value of availability performance. These are translated from the plant level to its subsystems, including the safety warning and response systems. A maintenance operations concept would have also accounted for maintaining the integrity of the warning and response system.
- Availability-centered engineering and management practices would

have been developed. They also automatically include issues germane to safety and hazards (AB1.8, Chapter 5).

The role of availability engineering and management goes far beyond the assessment of hazards. It also includes the control and management of safety and hazards.

A major dimension of process safety management is the mechanical integrity of plant equipment. This includes the safety system equipment. It also includes production equipment whose failures and maintenance results have hazardous ramifications.

Mechanical integrity is within the domain of availability engineering and management. Its requirements are an integral and normal part of the availability deliverables and tasks. Without it, neither this nor any other traditional deliverables will be effective.

TB11. Identify and Begin Procurement of Long-Lead-Time Delivery Items

Traditional Scope

The owner will have established a date for startup and commercial production. Initial scheduling will reveal that the procurement cycle for some equipment is too long for normal approaches. These items require that the process be expedited. Their acquisition may be initiated using preliminary equipment design, information, and specifications.

Pertinence of Availability Design

Availability engineering will contribute vital information to the procurement of long-lead-time items. Analysis will determine how critical these items are to availability performance. The design team will then provide availability parameters to procurement that will have been determined by preliminary availability modeling and data collection and analyses.

The availability engineering process will also participate in the preparation of purchase specifications. Areas of special interest would be:

- To communicate availability parameters.
- Availability design, testing, and evaluation requirements.
- Requirements associated with maintenance operation functions and systems.

- Requirements identified from the history of maintenance on similar equipment.
- Failure data in other plants.

TB12. Develop Equipment Specifications

Traditional Scope

The development of complete specifications began with the performance specifications (TB4). This process will have generated data sheets, which must now be formatted and attached to the performance data to form a complete specification.

Pertinence of Availability Design

Interdisciplinary participation. Fully-developed specifications reflect all pertinent design and management disciplines. Without availability engineering, the elements applicable to reliability and maintainability can only be superficially treated. There would be no disciplined approach to determine and include the various reliability and maintainability design details and requirements.

The critical difference between the two disciplines is that the traditional design process generates input/output type parameters associated with the production process. Availability engineering generates the parameters for time-between-failures and time-to-repair. It also identifies requirements for human factors and other "good practices."

Availability requirements in the purchase specifications. Purchase specification includes additional requirements associated with availability design and management. These include the following:

- Development of maintenance task detail. These are in terms of maintenance activities, time, parts, human resources, etc.
- Guidance for the preparation of procedures that are consistent with the maintenance operation concept.
- Data, information, and details to become part of the maintenance operation functions and systems.
- Data for life expectancy of equipment elements.

Maintenance task detail.   Maintenance task analysis and the development of procedures are costly but extremely valuable deliverables in the detailed design phase (AD1 and AD2 of Chapter 11). It may be very cost-effective to include the manufacturer in their development. The objective is to minimize the need to substantially revise traditionally provided technical materials.

The resulting efficiency and effectiveness are important for the following reasons:

- These work products are critical to computing maintenance and support resources, developing training programs, and assessing human factors and reliability.
- They are core elements of the living design and lifetime management of availability.
- Development of maintenance task analysis, procedures, and training programs is a significant cost. Any project management practice that reduces those costs is worthwhile since it provides an opportunity to use the owner's resources for other purposes.

TB13. Develop Bulk Material Specifications

Bulk material requirements and quantities are determined by materials takeoffs from the design drawings. The requirement at this point is to prepare the specifications for the process of procuring bulk materials for the construction phase.

Final Comments for Availability Engineering and Management in the Basic Design Phase

This chapter is the final discussion of the basic design phase. The traditional capital project deliverables produce a production process and plant that is capable of producing the product according to specification. Availability engineering will design it to most profitably achieve management's specified availability performance. The combined results are the optimal, maximally profitable plant productive capacity. Without availability engineering and management, the all-important productive capacity is only partially developed. But that is not all. There will also be considerable waste in plant maintenance cost. Costs for maintenance in a continuous process plant are often second only to raw material. They exceed others by order of magnitude.

From these chapters the reader may have made an observation. By describing what is required to design for availability, the weakness of the traditional work scope has become apparent.

This is an important observation. Many people involved in capital projects may have long been aware of the importance of the issues treated by availability engineering. Thus, they may recognize that the described availability deliverables are necessary to solve them. Anything short of that is only good intentions leading to unrealistic expectations.

The next four chapters describe availability engineering for the detailed design phase. Its deliverables are a continuation of the basic design phase. They focus on the detailed design of maintenance operations.

However, the detailed design-phase deliverables are dependent on all previous ones. This is because the business of maintenance operations is to deliver profitable plant availability performance. In that enterprise, it is no different than any other.

Enterprise business performance is contingent upon planning its functions to optimally serve its customers. This requires that the customer be analyzed and understood. In effect, the detailed depiction of the plant as the "customer" is the overall core product of the conceptual and detailed design phases.

Bibliography

Center for Chemical Process Safety. Guidelines for Hazard Evaluation Procedures. Institute of Chemical Engineers. New York, New York. 1989.

Center for Chemical Process Safety. Guidelines for Technical Management of Chemical Process Safety. American Institute of Chemical Engineers. New York, New York. 1989.

Lamb, Richard G. Availability Engineering and Management For Compliance With Mechanical Integrity in Process Safety Management. Technical paper to 1993 Petro-Safe Conference. Book 1, Volume II. Pennwell Conferences. Houston, Texas. January 1993.

Prett, David M. and Garcia, Carlos E. *Fundamental Process Control.* Butterworth-Heinemann. Boston, Mass. 1988.

Rase, Howard F., and Barrow, M.H. *Project Engineering of Process Plants.* John Wiley & Sons, Inc. New York. 1957.

Seminara, J.L. and Parsons, S.O. Human Factors Review of Power Plant Maintainability. Palo Alto, California. Electric Power Research Institute. 1981. EPRI NP-1567.

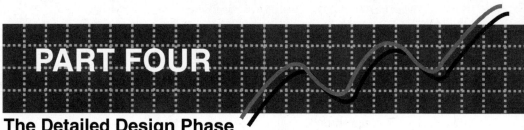

PART FOUR

The Detailed Design Phase

CHAPTER 11

Development of Maintenance Tasks and Procedures

Introduction

At the beginning of the detailed design phase, the configuration, size, and choice of equipment was established. At this point in the process, procurement of long lead-time items is well under way. The preparation of construction documents is now the basic objective. Meanwhile, activities are underway to procure and transport equipment and materials to the site in accordance with the construction schedule.

The purpose of availability engineering is to design the overall operations of availability and plant maintenance. Maintenance operations are designed to fulfill their fundamental business purpose, which is to deliver the availability performance and associated profitability the plant is capable of by virtue of its hard design.

Meanwhile, availability operations are designed to manage, drive, and assure the overall dynamic competency of the plant availability scheme. Achieving this purpose is a matter of organizational design for availability engineering and management. This includes participation in the change of the plant's hard design, development of and application of availability enhancing technology and methods, and incorporation of availability issues in the owner's business management and maintenance operations.

Flexibility of Availability Performance

At this point in the project, many aspects of the plant's characteristic reliability and maintainability have essentially become fixed. Chapter 1 introduced the top level factors that determine both. Some aspects are now fixed until some point in the future while others are not. The latter includes:

- Skill levels planned for maintenance tasks.
- The degree to which the maintenance tasks are forecasted and well developed.
- The quality of the procedures that detail and assign the maintenance task to relevant skill levels.
- Competence and effectiveness of the availability and maintenance supervision, planning, and management functions.
- The nature and quality of the training program with respect to maintenance tasks, procedures, supervision, etc.
- Opportunities for part interchangeability.
- The level of human and physical resources.
- Capacity and functionability of the maintenance and administrative support facilities.
- Durability of testing and maintenance equipment.

Opportunity to Be Cost-Effective

Like traditional plant design, the largest cost of availability engineering occurs in the detailed design phase. The biggest cost items are:

- Maintenance task analysis (AD1).
- Preparation of procedures (AD2).
- Development and implementation of the training program for maintenance personnel (AD6).

These are ultimately a part of maximally profitable plant functioning. Thus, availability engineering does not create new costs. Instead, these costs are incurred early in the plant's life.

The principle of the time value of money would suggest that this is not a financially attractive action. However, there are substantial benefits to this forward shift. They far outweigh the offsetting time value issue because:

- The requirements are developed as a holistic design. Therefore, they produce more effective and efficient results for maintenance operations.
- The development of these requirements costs much less than the costs to develop less effective alternate, interim, and ad hoc procedures.

- Possibilities will be created for plant design and planning that would not otherwise be the case. These include:
  - Human factors and reliability assessment are given empowering detail.
  - It would otherwise be impossible to compute human and material resource provisioning levels. This is a valuable benefit in terms of avoiding expenses against income and lost productivity of working capital resources.
  - Space requirements for facilities are able to be computed. Like resources, these could not otherwise be derived.
  - Maintenance operation functions can be better prepared for plant startup.
- It follows from the above that plant income will be greater. The plant will also reach its income potential sooner.

Organization of PART FOUR

The detailed design phase is presented in four chapters. The phase deliverables are grouped as follows:

- Development of maintenance task details and procedures (Chapter 11).
- Determination of human, material, and facility requirements for maintenance operations (Chapter 12).
- Analyses and design of availability and maintenance operation functions and organizational effectiveness (Chapter 13).
- Availability-centered review of traditional detailed design phase deliverables (Chapter 14).

The availability engineering deliverables are shown in Figure 11-1. They are coded as "AD." This is to denote an availability deliverable in the detailed design phase. The figure also shows that three deliverables will be discussed in this this chapter. They are:

- Maintenance task analysis (AD1).
- Needs analysis and design of maintenance, instructions, procedures, and manuals (AD2).
- Evaluation of equipment for maintainability (AD3).

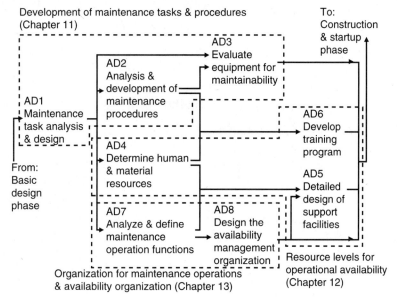

Development of maintenance tasks & procedures (Chapter 11)

Fig. 11–1 Availability engineering deliverables for the detailed design phase.

AD1. Maintenance Task Analysis

Definition and Purpose

Maintenance task analysis is critical to achieving specified availability performance. It is a bridge from the previous design phases to the deliverables of this phase.

The basic approach is simple. Each now-identified maintenance task is broken into its administrative, logistic, and active repair steps. All pertinent activity and resource details are attached to each step.

Thus, a powerful database is created. Many of the remaining phase deliverables draw upon it for their success. It is drawn upon and revised throughout the plant's producing life.

Degree of Rigor

Maintenance tasks are analyzed and then detailed for plant equipment, assemblies, and components. Not all maintenance tasks will be subjected to this process. FMECA will have first identified which failures are significant. These are candidates for methodical preparation.

The overall criteria for tasks to be subject to analysis are as follows:

- Where plant performance and safety are sensitive to the quality of the subject task.
- Failures that have significant repair requirements. These affect the calculation of human, materials, and facility resources.
- Maintenance tasks that make relatively complex demands on the organization. They require analysis to determine full support requirements.
- Tasks that require high competence in any of the universal maintenance activities. Recall that these activities are testing, troubleshooting, removing, repairing, replacing, calibrating, and servicing.

Tasks for Accomplishing the Deliverable for Maintenance Task Analysis

The repair requirements meeting the above criteria must be detailed as a maintenance task. This attention is warranted not only for efficiency and effectiveness. It is fundamental to the success of maintenance operations in its primary business of delivering the necessary availability performance the plant is capable of producing.

The tasks for performing the deliverable are flowcharted in Figure 11-2. The flowchart also shows sequences and iterations with tasks in parallel and previous deliverables.

AD1.1. Define, plan, and develop the maintenance task analysis system. Immense amounts of data and information will be acquired, developed, and organized. Therefore, the first requirement is to define, plan, and develop a format and system for this task detail.

The format should be in the form of a spreadsheet. Otherwise, there is no universal template. The format should fit the needs of the organization.

However, the contents are universal. The process for detailing maintenance tasks (AD1.3 and AD1.4) will introduce content elements.

Maintenance task analysis should be developed in an electronic system. The reasons are as follows:

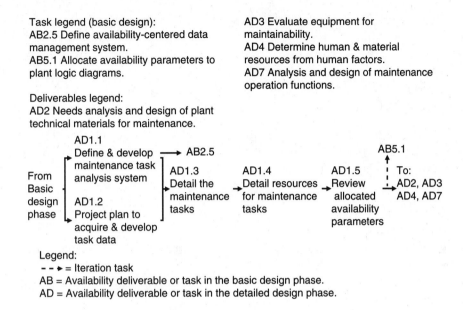

Task legend (basic design):
AB2.5 Define availability-centered data management system.
AB5.1 Allocate availability parameters to plant logic diagrams.

Deliverables legend:
AD2 Needs analysis and design of plant technical materials for maintenance.

AD3 Evaluate equipment for maintainability.
AD4 Determine human & material resources from human factors.
AD7 Analysis and design of maintenance operation functions.

AD1.1
Define & develop maintenance task analysis system

From Basic design phase

AD1.2
Project plan to acquire & develop task data

AB2.5

AD1.3
Detail the maintenance tasks

AD1.4
Detail resources for maintenance tasks

AD1.5
Review allocated availability parameters

AB5.1

To:
AD2, AD3
AD4, AD7

Legend:
- - ► = Iteration task
AB = Availability deliverable or task in the basic design phase.
AD = Availability deliverable or task in the detailed design phase.

Fig. 11–2 Tasks for maintenance task analyses (AD1 of Figure 11-1).

- To provide a means to both collect, extract, apply, and refine massive data and information. These events are connected with the analysis process and the needs of other deliverables.
- To serve corporate, operating company, and plant management functions throughout the facility's producing lifetime. This purpose goes beyond plant maintenance operations.
- To provide the ability to trace the consequences of plant changes and improvements throughout the availability scheme.

The data and analysis system most likely will have many uses. It will probably include or integrate components of other maintenance operations systems. Examples are equipment information, work package, and materials systems.

This task is partly one of computer systems integration. However, it is not the first time the issue has been addressed. The conceptual design phase availability concept deliverable included a task for formulating an integrated computer system concept (AC2.10, Chapter 3). Maintenance task analysis as a subsystem was included in the concept.

Figure 11-2 shows that the result of this task will flow to a task in the basic design phase to define the overall data management system (AB2.5, Chapter 5). The overall scheme will be refined by this input.

AD1.2. Plan and organize for data acquisition and development.   The data and information requirements of this deliverable are a substantial endeavor. The payoff is immense in the short and long term. But as a project activity, it is a lump in the proverbial snake's belly.

Consequently, this task is to plan and organize for data and information. The basic planning issues are as follows:

- Sources and quality of data and information. This includes data to be developed by previous and subsequent deliverables. It is also important to plan for data sources over the plant's lifetime.
- Anticipated difficulties in the acquisition and development of data.
- The approach and plan for data acquisition and development.

A possible approach is to assign the problem to a business that manages such work as a specialty. One benefit is that they have already developed electronic systems for maintenance task analysis.

Project management should evaluate these businesses for their database and management processes for searching out and assigning specialized contract personnel to each maintenance task. Thus, their value is to assign personnel according to experience with plant items and operating conditions.

Another approach is for the project team to collaborate with equipment manufacturers. They may be initial contributors or they may provide assistance on an as-needed basis. Possibilities include:

- Historically, the manufacturer provides maintenance procedures. Therefore, they can provide the detail for maintenance task steps in an appropriate format.
- They can provide the detail of parts and other information associated with each maintenance task step.
- They can provide much of the necessary data and information elements associated with the subject equipment.
- In many cases, they can provide the required detail in electronic form. This can be easily transferred to the various availability analysis and design models.

Therefore, project management may wish to establish manufacturer requirements beyond the traditional approaches. A possible scenario is for the design team to supply the manufacturer with certain detail. Possibilities include the maintenance operation concept, skill level decisions, parameters for time-to-maintain, etc. The manufacturer can then better provide an identification of all maintenance tasks, maintenance task steps, equipment and parts detail, etc.

AD1.3. Detail the maintenance tasks. Maintenance task analysis involves two types of detail. They are task detail and associated resources. The former is produced by the following subtasks (Figure 11-3):

AD1.3.1. Identify the subject plant component. The gathered information should include include the following:

- Part, assembly, or component,
- Next higher system or assembly,
- Subsystem.

AD1.3.2. Identify task location and working environment. The expected working environment should be included in this detail.

AD1.3.3. Describe the requiremer t for the task and its result. The reason the task is required and its result are described. Examples are to restore a function, return a function to full capacity, etc. In the basic design phase (AB8, Chapter 8) FMECA first developed this information. It is now extended to this task.

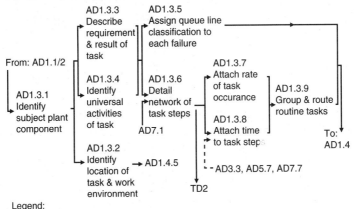

Task legend:
AD1.1/2 Define maintenance task analysis systems and plan for data development.
AD1.4 Detail resources for each maintenance task and its steps.
AD1.4.5 Identify facility & associated equipment requirements.
AD3.3 Evalulate equipment for maintainability from human factors.

AD5.7 Incorporate the results of facility analysis in maintenance task analysis.
AD7.1 Develop process & linkage diagrams for maintenance operations.

Deliverable legend (traditional):
TD2. Detailed safety and hazards analysis.

Legend:
- - ► = Iteration task
AB = Availability deliverable or task in the basic design phase.
AD = Availability deliverable or task in the detailed design phase.

Fig. 11–3 Subtasks to detail the maintenance tasks (AD1.3 of Figure 11-2).

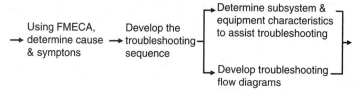

Fig. 11–4 Steps for developing a troubleshooting process for each failure.

AD1.3.4. Identify the universal activities of the task. All maintenance tasks involve one or all universal activities. These were introduced earlier as testing, troubleshooting, removing, repairing, replacing, adjusting, and servicing. This subtask is to identify which are involved in the subject maintenance task.

Troubleshooting procedures are a special maintenance activity. They are designed as part of maintenance task analysis. It is worthwhile to take a closer look at the steps involved. The process also shows another purpose of FMECA in the basic design phase.

Troubleshooting is triggered by an operation, inspection, or failure event. An assumption is made that an element is faulty. Maintenance is notified. The maintenance person, in conjunction with operations personnel, then determines if maintenance is necessary by inspection or by causing the system to repeat the fault. The symptoms are identified. A logical process is then followed to find the source or cause.

One objective of designing the troubleshooting process is to create the detail that will avoid trial and error approaches. Trial and error is costly in terms of time, resources, and, therefore, plant availability performance.

The steps for developing the troubleshooting process for each failure are as follows (Figure 11-4):

1. Using the previously developed FMECA, determine the cause of each failure mode. The failure is expressed as a symptom. In essence, FMECA is a proactive root cause analysis.
2. For each failure mode or symptom, develop the troubleshooting sequence. The FMECA also provides information for this step.
3. Determine the necessary subsystem and equipment features that would better allow the troubleshooting sequence to progress. These include access points, meters, test points, etc. These are incorporated in equipment design. Such details may be relatively flexible even late in the equipment manufacturing phase if they do not entail fundamental changes in its design.
4. Develop flow diagrams of the troubleshooting process.

AD1.3.5. Assign a queue line classification to each failure. Each failure is classified for relative placement in line for service. The previous phase provides considerable input for this determination. FMECA will have determined and ranked the consequences of the failure. The results are extended to this task.

AD1.3.6. Detail the network of steps for each maintenance task. The maintenance task analysis system should include the capacity to detail each task as a network of steps. The detail should include active repair, logistic, and administrative steps. As a result, the detail is a part of the job plan for each failure. Later tasks will complete the job plan detail.

This is the network scheduling concept applied to each maintenance task. The task steps are its elements. The level of resolution is determined by a simple criterion: It must be possible for resources and time requirements to be distinguished and attached to a maintenance step. Thus, job standards are one consequence of such detail.

Figure 11-3 shows that the task step detail will be refined by a later task. The deliverable for detailing maintenance operations includes the task of defining the process and linkages for administering and providing resources to the maintenance task (AD7.1, Chapter 13).

The figure also shows that the results flow to the detailed safety and hazards assessment (TD2, Chapter 14). The assessment will evaluate the hazardous consequences of failing to do a step or of doing it poorly.

AD1.3.7. Attach the rate of occurrence to each task step. The statistical frequency of task occurrence is directly connected to the expected frequency of a necessary repair. This information is attached to each step of the subject task. This is made possible by the databases developed in the basic design phase. They are the result of developing reliability data and parameters for availability modeling.

AD1.3.8. Attach the time for each task and step. Required time is attached to the steps of each maintenance task. The result is the expected total elapsed time to return each failed item to service.

Total elapsed time is the composite of active repair, logistic, and administrative times. At the beginning, time detail is limited to active repair steps.

The latter two evolve from decisions made in later design deliverables. First, logistic time will come from the analyses of resource levels and support facilities (AD4 and AD5). Second, a later task will ultimately synthesize the findings of service and consumable resources and facilities analysis as a system of resources (AD5.6). Figure 11-3 shows that the results (AD5.7) will flow to this subtask.

Time for the administrative elements will evolve from a later deliverable to analyze and design the maintenance operation functions and systems (AD7, Chapter 13). A network of processes is detailed (AD7.1). Administrative times is then estimated for each process activity (AD7.7).

Figure 11-3 shows an additional refining input. A task of a later deliverable will assess equipment for maintainability and human factors (AD3.3). At issue is the ease or difficulty of actually performing the active repair steps. The findings may lead to a revision of required active repair steps and their time to complete.

AD1.3.9. Group and route routine tasks.   Routine inspection, service, and adjustment are planned as grouped activities. The result is a regular route of tasks. The route is a function of location, required skills, and which entities the work is assigned to. Routing decisions are also subject to physical restrictions. This includes barriers and other separating aspects of the plant's layout.

AD1.4. Detail the resources for each maintenance task and its steps.   The previous task analyzed and detailed the maintenance tasks and their steps. The process is now extended to identify in detail the associated human and material resources, instructions and procedures, training, facilities and other elements of those steps.

The subtasks for defining the resource elements for each step in the maintenance tasks are flowcharted in Figure 11-5. They are as follows:

AD1.4.1. Identify human skill requirements for each task.   The human resource requirements must be identified for each task step. This includes skills and the source service entity. The latter was first established by the maintenance operation concept (AC3.1, Chapter 3) in the conceptual design phase.

AD1.4.2. Identify all parts.   For each task step, identify all replacement parts, components, and consumables. This detail is extended to related data such as quantity, part number, nomenclature, etc.

This requirement tests the procurement process and the data it has specified in the purchase specifications. The maintenance task analysis accordingly provides a tool for confirming that all necessary data has been identified, specified, and gathered by the procurement processes.

The resulting data and information fields are the foundation for parts cataloging systems and functions. These are crucial for cost-effective parts planning and stocking.

AD1.4.3. Identify standard and special tools. Tools should be identified as standard or special requirements. The detail is extended to include all relevant identifying administrative and management information.

AD1.4.4. Identify testing and handling equipment. Testing and handling equipment must be identified for each task step. This data includes quantity, nomenclature, and source. Attached to this data field is additional data and information such as where the item is to be used, the working environment, testing parameters and tolerances, calibration requirements, whether the subject item is off-the-shelf or engineered, etc.

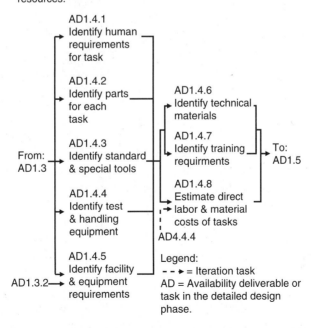

Task legend:
AD1.3 Detail the maintenance tasks.
AD1.3.2 Identify task location & working environment.
AD1.5 Review & revise availability parameters allocated to plant logic diagrams.
AD4.4.4 Compute economic order quantities for consumable resources.

Fig. 11–5 Subtasks of detailing maintenance task resources (AD1.4 of Figure 11-2).

AD1.4.5. Identify facility and associated equipment requirements. The location of each task was identified in a previous subtask (AD1.3.2). The details include required facilities, space, capital equipment, storage, etc.

AD1.4.6. Identify technical materials. Identify the technical documents required for each task step. This data field should indicate the distinction for instructions versus procedures. Such distinction is associated with the earlier identified skill level.

AD1.4.7. Identify training requirements. Training requirements must be identified for each maintenance step. Information includes modules for specific skills and training to achieve and maintain proficiency.

AD1.4.8. Estimate direct costs of labor and material. Direct labor and material costs must be estimated for each maintenance task. This is compared to indirect or overhead costs. Another financial facet is the asset balances associated with the direct costs. They will be determined in later tasks (AD4.3.6 and AD4.4.7, Chapter 12).

The findings for direct costs flow via the next task to upgrade the cost estimates initially developed in the basic design phase. They were the initial estimates of life cycle costs for plant items. From there the revised estimates will flow to the life-cycle cost subsystem of the financial model.

Figure 11-5 shows that the costs for material resources may be revised by the later analysis of economic order quantities (AD4.4.4, Chapter 12). This is because a purchase discount may be revealed as a function of the stock replenishment order size.

AD1.5. Review and revise the allocated reliability, maintainability, and economic parameters. It is now necessary to compare the results of the maintenance task analysis with the availability parameters previously developed in the basic design phase. They were developed as the basis for initially calculating plant availability.

There may now be differences in estimated time-to-return-to-service and associated economics. Dissecting the maintenance task provides a more rigorous assessment. Therefore, it may be necessary to reconcile the previously allocated parameters with these findings.

There are two stages of reconciliation. One is possible when active repair elements are detailed and direct costs are attached. The other is possible when resources and organizational effectiveness are developed and costed by later deliverables.

The first stage occurs as active repair data leads to the review of expected achievable availability (A_a) and its economics. Achieved availability was defined as the expected performance when logistic and administrative times are not included. Thus, the first stage refines this dimension of forecasted performance. The second stage occurs as the maintenance task analysis accumulates the detail necessary to attach time for logistic and administrative elements in the network of task steps (AD1.3.6.).

Figure 11-5 shows that the results of both stages will iterate to the basic design phase through the task for allocating availability parameters to plant logic diagrams (AB5.1, Chapter 6). From there, the computation and optimization of availability performance is refined.

That task, in turn, refines two previous ones: the analysis of reliability and maintainability data (AB3.4, Chapter 6) and the development of life cycle costs (AB3.5).

Plant and Task Numbering System

An equipment-based list, numbering, and coding scheme are usually developed in the traditional design steps. The availability engineering discipline has an immense vested interest in their development.

This is apparent by the above description of the maintenance task analysis in the previous section. Most of its elements are connected through a carefully defined equipment-based numbering and nomenclature system (AD1.1).

AD2. Needs Analysis and Design of Maintenance Instructions, Procedures, and Manuals

Definition and Purpose

This deliverable is concerned with the needs analysis and development of instructions, procedures, and associated manuals. The key words are needs analysis. Instruction and procedure manuals are often viewed as something provided by manufacturers. From there, they are placed on a shelf ready for use. However, it is not a safe assumption that these materials will actually fit the plant's maintenance operation concept.

Importance and Implication of Needs Analysis

Rigorous needs analysis and development of procedures should be proactive preparation. Otherwise, maintenance task analysis and procedures development may be accomplished under pressure in the field on somebody's knees. The lesser the skill level of those involved in forming such ad hoc maintenance solutions, the greater the potential of an unfortunate result.

Association of procedures to accidents. The Occupational Safety and Health Administration (OSHA) has found that poor procedures have historically contributed greatly to accidents. It follows that this is a reflection of commonly observed weakness in instructions and procedures. Examples are as follows:

- A mismatch between the content of the procedure and the skill level of the user.
- The instructions are written by the manufacturer to cover multiple models of equipment. They are not written exclusively for the subject model in the plant.
- They include diagrams that are poorly organized and developed.
- They suffer from the lack of an effective organization to test, evaluate, and manage revision processes. Recall that the organizational processes for change, improvement, data, and baseline management were developed in the basic design phase (AB2). One purpose is to avoid this problem.

The consequences are suggested in a study of 988 license event reports [Brune and Weinstein, 1980] submitted to the nuclear Regulation Commission. In 751 reported maintenance related cases, human error was found to at least be partially the consequence of marginal procedures.

Modern process safety management approaches look .beyond such error. Their philosophy is that human error is ultimately traceable to a failure in some management practice.

Procedures provided by manufacturers. The above points have implications. The most obvious is that manuals provided by manufacturers should be initially considered as only input or draft materials for the ultimate technical manuals. An assessment against maintenance operation detail will determine whether they are acceptable for field use. A basic test of fitness is whether they are written to fit the plant, skill levels, common formats, and defined library system.

Continued testing and evaluation. There is another implication. The most important technical materials must be subjected to testing and evaluation initially and periodically throughout the plant's life. The management systems for such a purpose must also be configured to effectively and efficiently revise the manuals. This will be necessary as the plant's mission and availability scheme changes over time.

Tasks for the Deliverable of Analyzing and Designing Technical Materials

The tasks of performing the deliverable are described in the following sections. They are flowcharted in Figure 11-6. The flowchart also shows the sequences and iterations with tasks in parallel and previous deliverables.

AD2.1. Define and plan the library and system for technical materials. Instruction, procedures, and manuals constitute a large body of information. How they are managed is an important design issue. This includes the processes for document control such as dating, control of copies, assuring the use of updated procedures, etc. Therefore, this task is to assess and define such aspects.

Figure 11-6 shows that the results iterate to the basic design phase task to define the availability-centered data management system (AB2.5, Chapter 4). It also shows that a task to develop the training programs may affect the decisions of this task. Technical materials may be selected to also serve as training materials (AD6.3, Chapter 12). Such decisions may impact the conclusions of this task.

AD2.2. Plan the project to acquire and develop technical materials. It is important to plan for the acquisition and development of technical materials. Project plans must be concerned with the following:

- Possible sources and producers and their roles.
- Expected initial conditions or the nature of the materials.
- Requirements for their development or refinement.
- Strategies for their acquisition and development.
- Schedule for their acquisition and development.

An important project planning decision is who will do the actual work. One possibility is to assign responsibility for the final work product to the person who will have maintenance responsibility for the facility. A possibility for the actual development tasks is to distribute some of the workload to the manufacturer since they have a historical role in the provision of technical materials.

The permanent staffing of specialized personnel for this deliverable is generally not economically feasible for the owner or the main project contractor. An alternate approach is to assign the requirement to companies that specialize in producing technical documents.

Such organizations are characterized by their human resource management tools and approaches. This is because large numbers of contract personnel are crucial for a timely result. They are selected for their specialized knowledge and experience with each item of equipment.

The most attractive and likely approach is the integration of manufacturers and a service organization. The former should be assigned the task of producing the technical materials according to defined practices. The service organization should be responsible for management across manufacturers. They will also augment and complement the manufacturers' capabilities and contribution.

AD2.3. Define the organization and staffing for technical materials management.   Management of technical materials is a significant organizational function. This task is to formulate an organization design for that pur-

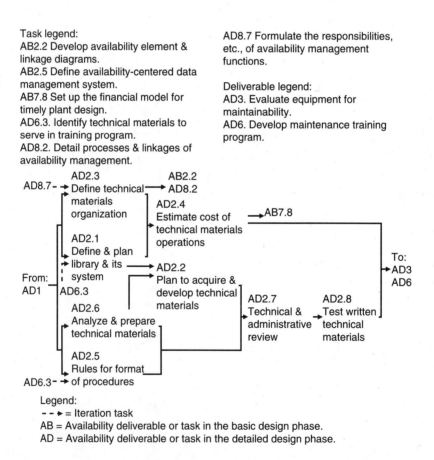

Fig. 11–6 Tasks for needs analysis and design of maintenance instructions, procedures, and manuals (AD2 of Figure 11-1).

pose. The scheme includes all related processes and their linkages. These are processes for storage, access, critical feedback, improvement, change, acquisition, etc.

The approach will not be described here. It is essentially the same as the approach described for designing the overall availability management organization (AD8).

The result of this task is preliminary until refined and finalized by that deliverable. Consequently, the scheme of this deliverable will be synthesized with those of other functions of the overall availability engineering and management organization (AB8.2, Chapter 13). The results of this synthesis (AD8.7, Chapter 13) will flow back to this task.

The results of this task will also flow to the one for detailing linkages and processes as part of the overall availability-centered improvement, change, and data management systems (AB2.2, Chapter 6).

AD2.4. Estimate the costs of the technical materials operation. The entire scheme of technical materials management is defined by the previous tasks. This task determines the cost of that scheme.

Figure 11-6 shows that this task will supply information to the plant financial model (AB7.8, Chapter 7). Estimated management and staff expenses for each accounting period will flow to the income statement subsystem. These expenses will generate a need for associated working capital balances.

Asset balances for the cost of acquired documents and systems. These will flow to the balance sheet subsystem of the financial mode. How they will be converted to the income statement subsystem must be decided.

These costs and balances are not directly attached to maintenance tasks. Thus, they will not appear as economic parameters assigned to the plant elements. Nor will they pass through the life-cycle cost system in the financial model. However, they are a significant financial elements of maintenance operations.

AD2.5. Establish rules and procedures for developing technical materials. The development of technical materials may have begun in the basic design phase. Requirements and rules for the contents of technical materials may have been established by the deliverable of developing "practices control" documents (AB1). If this was not the case, these tasks should be done at this time. The results should be incorporated in the library of practices documents.

The work product is a set of very specific requirements. The objective is consistency between all technical materials. This is important since equipment is supplied by many manufacturers.

Figure 11-6 shows that the formulated rules may be affected by the needs of the training programs. Some materials may later be identified to serve a training purpose (AD6.3, Chapter 12).

The fundamental requirements for "good practices" are as follows:

- A format and table of contents that the manufacturer, designer, and owner will follow for all technical materials. Format definition should be rigorously detailed. For example, the format may specify that the text serve as a checklist for task steps. This is necessary when an omission can undermine reliability or safety. The format may specify that a space be provided to record important field information.
- Sections included for the theory and principles of equipment features and operations.
- Identification of tools, parts, and equipment.
- Instruction segregated by the universal maintenance activities. It was mentioned that these are inspecting, troubleshooting, removing, replacing, repairing, etc.
- An index of requirements and principles.
- Troubleshooting charts. These must minimize the use of "contact manufacturer" as a solution.
- Instructions and procedures should be limited to one item of equipment.
- Documents should be written for the skill levels of the personnel planned to use the instruction or procedure. The perspective of this issue as one skill level may be narrow. The issue may instead be what skill levels may possibly come to use a procedure. Thus, it may be necessary to write the document to serve a range of skill levels. For example, it may be formatted so that a highly skilled person need only read the headings and use underlined key data. Alternately, the lesser skilled user may need to use headings and the full text to accomplish the same assignment.

  Beyond the obvious, the match between skill level and technical materials is very important. A highly skilled worker may not read materials written in great detail. Excess detail may cause the task to be done by memory, intuition, and experience. A much lesser skilled person cannot function competently with only instructions. They require detailed procedures. Thus, a mismatch can make the the diligently developed technical materials almost meaningless.

- Rules for graphics with regard to the information and actual instructions they contain; the scheme for relating text to the diagram details; and location with respect to the text.

- Location of precautions and operating limits with regard to the applicable steps.
- Identification of specific documents that are either input or output for the maintenance task.
- Principles for writing instructions and procedures. This may include rules for the number of actions in each step; common usage of terminology and indexing; phrasing rules and patterns such as sequences of special tools, verbs, objects, locations, and movements; and the length of sentences.
- Size, binding, and durability that are suited to use in the field.

AD2.6. Analysis and preparation of technical materials. The steps for analysis and preparation are applicable regardless of source. The subtasks are as follows (Figure 11-7):

AD2.6.1. Extract requirements from maintenance task analysis. Maintenance task analysis (AD1) has developed information vital to the analysis and preparation of technical materials. Requirements are extracted and formatted as follows:

1. Extract a matrix of equipment items assemblies and components versus associated maintenance activities. The latter includes repairing, replacing, inspecting etc.
2. At the points of intersection, extract an identification of the type of required technical materials. These too were established by the maintenance task analysis for each item and its associated maintenance activities.

Task legend:
AD2.2 Plan the project to acquire & develop technical materials.
AD2.7 Technical & administrative review of technical materials.
AD4.3.2/3 Determine service & consumable resource scheme, levels & costs.

Deliverable legend:
AD1 Maintenance task design & analysis

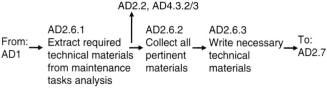

Legend:
AD = Availability deliverable or task in the detailed design phase.

Fig. 11–7 Subtasks to analyze and prepare instructions and procedures (AD2.6 of Figure 11-6).

Figure 11-6 shows that this detail flows to tasks in the deliverable to compute resources. They are to compute the frequency and time of application (AD4.3.2/3, Chapter 12).

AD2.6.2. Collect all pertinent materials. Materials provided by a manufacturer should be considered as draft materials rather than a final document, until proven otherwise.

AD2.6.3. Write technical materials. Write the instructions, procedures, and manuals according to guidelines for good practices (AD2.5).

AD2.7. Conduct a technical and administration review. The written instructions, procedures, and manuals must be subjected to a technical and administrative review. The objective is to evaluate whether each is:

Written correctly.
Technically accurate.
Consistent with plant and corporate documents and standards.
Safe to perform.
Readable and understandable at the skill level of field personnel, void of irrelevant and extraneous information.

AD2.8. Test and evaluate written technical materials. The objective is to confirm that the procedure will actually be effective in the field. However, this is not a one-time process. It is done initially during startup or earlier, then periodically evaluated throughout the plant's remaining life as experience is gained. This should also be done when the results of change and improvement are traced to the subject materials.

The testing and evaluation subtasks flowcharted in Figure 11-8 are as follows:

AD2.8.1. Plan for testing and evaluation. Prepare written plans for talk-through, mock-up, or actual field applications.

AD2.8.2. Perform testing and evaluation. Appropriate personnel are used to perform the testing and evaluation. Care must be taken to assure that observers do not somehow render the test invalid or inaccurate.

AD2.8.3. Post-test review. Conduct a post-test review with all participants.

AD2.8.4. Revise procedure. Undertake a revision cycle in response to findings.

AD2.8.5. Document findings. Document the findings in accordance with the procedures established to manage revisions.

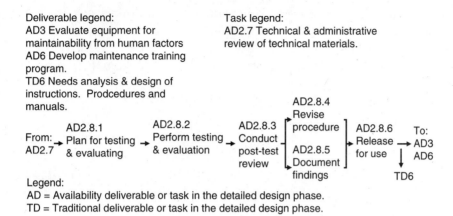

Deliverable legend:
AD3 Evaluate equipment for maintainability from human factors
AD6 Develop maintenance training program.
TD6 Needs analysis & design of instructions. Prodcedures and manuals.

Task legend:
AD2.7 Technical & administrative review of technical materials.

Legend:
AD = Availability deliverable or task in the detailed design phase.
TD = Traditional deliverable or task in the detailed design phase.

Fig. 11–8 Subtasks of testing and evaluating instructions and procedures (AD2.8 of Figure 11-6)

AD2.8.6. Release for use. Release the technical materials for unrestricted use. Ideally, the traditional project process was to develop technical materials for plant maintenance (TD6, Chapter 14). Accordingly, the released technical materials should fulfill that vision.

AD3. Evaluate Equipment for Maintainability from Human Factors Perspective

Definition and Purpose

The maintenance analysis (AD1) task and subsequently developed technical materials (AD2) assume that the equipment has been designed by previous deliverables for maximum maintainability. This deliverable is to test that assumption. The findings may reveal the need to revise the task details and procedures. In turn, they may lead to revising the allocated availability parameters (AB5) and availability scheme (AB10).

Human Factors Affecting Equipment

Human factors affecting equipment have the following dimensions.

- Maintainability of equipment.
- Protecting people from equipment and equipment from people.

The discussion that follows is intended to be an overview presentation of the human factors for maintainability and protection. These are the subject of many books. However, the discussion will eventually identify the major elements of equipment design for maintainability and human factors.

Possible Flexibility of Plant Hard Design

It is no longer widely possible to design equipment for human factors at this phase in plant design. However, it is expected that major shortcomings should not be discovered during the execution of this deriverable. The availability-centered "good practices" control documents (AB1) will have already detailed the rules for human factors. Project reviews will have confirmed that these practices were incorporated in equipment design, specifications, and manufacture.

Therefore, the primary objective of this deliverable is to evaluate the maintenance task detail and associated technical manuals. They are tested against the established plant and equipment design.

In the "grass roots" plant, maintenance task analysis and procedures are typically not available for the design of equipment for maintainability. This is because of the natural timing of information in the typical capital project process. Therefore, a reactive approach to evaluating human factors against specific maintenance activity steps is unavoidable.

It may still be possible to make design changes. These will probably be limited to small nonfixed aspects such as fasteners, handles, test points, etc. Changes may also be possible for equipment that has not progressed too far in its manufacture. This suggests that involving the manufacturer in the maintenance task analysis (AD1) may increase the opportunity to review and change design.

Therefore, the greatest area of change in response to human factors rests within the maintenance tasks details. If that is not possible, the second choice is to change the human and physical resource levels (AD4). This will compensate for the working conditions that are found to be less conducive than expected.

Tasks for the Deliverable of Assessing Equipment for Maintainability from Human Factors

The tasks for performing the deliverable are described below. They are flowcharted in Figure 11-9. The flowchart also shows the sequences and iterations with tasks in parallel and previous deliverables.

AD3.1. Determine maintenance tasks and issues for evaluation.   There are many equipment items, subassemblies, and components in a plant. For

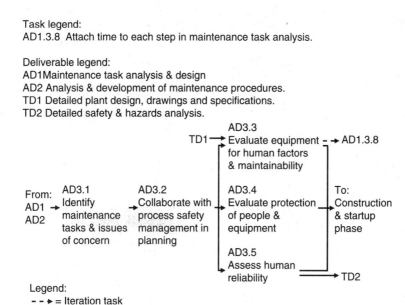

Task legend:
AD1.3.8 Attach time to each step in maintenance task analysis.

Deliverable legend:
AD1Maintenance task analysis & design
AD2 Analysis & development of maintenance procedures.
TD1 Detailed plant design, drawings and specifications.
TD2 Detailed safety & hazards analysis.

Legend:
- - ➤ = Iteration task
AD = Availability deliverable or task in the detailed design phase.
TD = Traditional deliverable or task in the detailed design phase.

Fig. 11–9 Tasks for assessing equipment for maintainability (AD3 of Figure 11-1).

each, there are many checkpoints for maintainability and human factors. This makes it unfeasible to undertake a general assessment of plant layout for maintainability (AB11). This is also true for the assessment of plant equipment for maintainability with respect to human factors.

The solution is still to focus analysis from the vantage of important maintenance tasks and their steps. These are associated with failures that are frequent and costly. They are also applicable to maintenance that, if poorly done, can lead to an undesirable consequence. One such consequence is reduced equipment reliability.

These cases were determined earlier (AB11.1, Chapter 10). The objective was to assess the plant layout. This task includes retrieving and applying the resulting list.

AD3.2. Collaborate with production process safety management in hazard assessment. Evaluation of human factors is in essence to test the task details and procedures against field conditions created by hard design. Some are assessed because of their relationship to an identified hazard. A catastrophic event resulting from a hazard is a specialized loss and cost of availability performance. Both are to be avoided.

Accordingly, this task is to collaborate with related process safety management initiatives. The objective is to determine the following:

- Maintenance tasks of concern to safety management. These were revealed by FMECA (AD8) and other methods.
- Safety management's associated scheme for managing and controlling the hazard.
- Safety management's desired and necessary role in this deliverable and its tasks.
- Common approaches to both availability and hazards analysis.

AD3.3. Evaluate equipment for human factors and maintainability. Equipment is evaluated for maintainability from human factors. As mentioned, this is guided by the relative importance of individual plant maintenance tasks.

The points for evaluation are as follows:

- Accessibility to or surveillability of all points on equipment subject to inspection, testing, servicing, adjustment, removal, and replacement. Accessibility is also concerned with whether there is potential for damage to adjacent or connected equipment, assemblies, and components, and whether access is away from dangerous conditions. It should also make allowances for clothing and protective gear worn in the plant and its environments.
- Standardization as a function of common manufacturers, equipment that can use standard hardware and hardware that is available as a commodity, maintenance actions that use standard tools, and general purpose testing and handling equipment. The possibilities for standardization are explored by the parts, tools, and maintenance equipment cataloging capability made possible by the content of the maintenance task analysis process.
- Modularization for electrical, electronic, and instrumentation and process control equipment will be concerned with: clearly labeled modules and sockets, ease of making connections, size of the subject equipment, proper inclusion of handles, the ability to check and adjust modules as individual units rather than having to test signals from multiple modules, and components that can be removed without disturbing other modules.
- Layout, mounting, and packaging of electrical, electronic, and instrumentation and process control components. This facet of evaluation is concerned with the way connections are made. Issues are ease and proper connection and the manner in which components are secured.
- Connectors for ease, simplicity, and minimum number of required connections.

- Fasteners in terms of number of fasteners; ability to be manipulated by one hand and whether they are large enough to be manipulated easily; whether standard tools are required; and clearance around the fastener.
- Handling with respect to rules for handles and lifting eyes.
- Access to test points for plant equipment. The maintenance task analysis deliverable (AD1.3.4) included the development of the troubleshooting process. The process identified the need for specific test points as part of the analysis. Equipment must be assessed to assure that these test points have been created, are appropriate, are located close to the problem, and are clearly marked. Location is especially important when external equipment is to be used.
- Labeling is important for avoiding delays in search of equipment, for preventing items from being missed during routine surveillance and servicing, for preventing repair to the wrong item, and for preventing injuries. The design issues for labeling include placement, orientation, contrast, size, style, terminology, abbreviation, symbols, durability, etc.

The elements of maintainability have been detailed and delegated to the traditional design tasks. As mentioned, the detail was developed as part of the availability-centered "good practices" documents for plant design (AB1).

Figure 11-9 shows that the findings of the task will iterate with the maintenance task analysis. They will be used to review the time allowed for each maintenance task step (AD1.3.8) to account for field conditions. Refinements will flow back back to the availability parameters allocated for plant logic (AB5.1). This feedback will be processed by another maintenance task analysis task (AD1.5).

The figure also shows that the findings will flow to the traditional plant design process. They will be incorporated in the tasks for detailed design, drawings, and specifications (TD1, Chapter 14).

AD3.4. Evaluate the protection of people and equipment.   Human factors for the protection of people and equipment must be evaluated. There are five possible strategies. In order of preference, they are as follows:

- Design for protection with approaches such as interlocks.
- Remove people from hazards. Strategies are remote test points, controls, closed circuit television, robotics, etc.
- Provide barriers that do not impede maintenance tasks beyond what is necessary. If obstructive, they may be removed and their purpose defeated.
- Provide warning labels. Maintenance task analysis will help identify the best location for warning labels.
- Establish training programs.

Many of the requirements for equipment maintainability and human factors are highly relevant to the protection of people and equipment. This is especially so for accessibility and handling.

AD3.5. Assess human reliability. Process safety management may plan to undertake a human reliability assessment. Different methods may be used. They will reflect the approach chosen to detailed safety and hazards assessment in the traditional project process (TD2, Chapter 14). Regardless of choice, the general approach to this task is as follows:

1. For each maintenance task step, decide if a potential hazard will be created if the task is not done correctly.
2. Use event-tree approaches to trace the progression of possibilities from the subject maintenance task step. A foundation for this and the previous step is provided by the existing FMECA. It revealed where failures have consequences with respect to a hazard becoming a catastrophic event. This task is to build on that analysis.
3. Determine additional task steps or administrative procedures for preventing or breaking the possible chain of initial and interim events.

Figure 11-9 shows that results of this task flow to the traditional design process. It becomes part of the traditional deliverable for a detailed safety and hazard assessment (TD2, Chapter 14).

Bibiliography

Blanchard, Benjamin S. *Logistics Engineering and Management.* 4th ed. Prentice Hall. Englewood Cliffs, N.J. 1991.

Brune, R.L. and Weinstein, M. Procedures Evaluation Checklist for Maintenance, Test and calibration Procedures.

Center for Chemical Process Safety. Guidelines for Technical Management of Chemical Process Safety. American Institute of Chemical Engineers. New York. 1989.

Pack, R.W., Seminara, J.L., Shewbridge, E.G. and Gonzalez, W.R. Human Engineering Design Guidelines for Maintainability. General Physics Corporation. Atlanta, Georgia. EPRI NP-4350. 1985.

Sanders, Mark S. and McCormick, Ernest J. *Human Factors in Engineering and Design.* 7th ed. McGraw Hill. New York. 1993.

Seminara, J.L. and Parsons, S.O. Human Factors Review of Power Plant Maintainability. Palo Alto, California. Electric Power Research Institute. 1981. EPRI NP-1567.

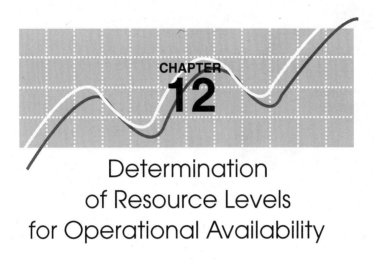

CHAPTER

12

Determination
of Resource Levels
for Operational Availability

Introduction

The previous chapter described the deliverables and tasks for designing and evaluating plant maintenance tasks. Because of that work, it is now possible to determine the levels of human, material, and facility resources required to achieve management's specified operational availability (A_o).

Recall that achievable availability (A_a) assumes a perfect maintenance environment. Perfection means that resources are always available and that there is maximum possible organizational effectiveness. When the decision for actual levels of resources and effectiveness are incorporated, the result is operational availability (A_o).

Figure 9-4 showed that operational availability is a level of performance below achievable availability. Its theoretical maximum is just below achievable availability (A_a). Therefore, operational availability is the bottom line of performance.

Management will have made strategic decisions for the relative position between operational and achievable availability. This was first addressed with the availability concept (AC2). It was later refined as the availability scheme was ultimately optimized (AB10, Chapter 9) in the basic design phase. It is now time to compute the resources associated with those strategic decisions.

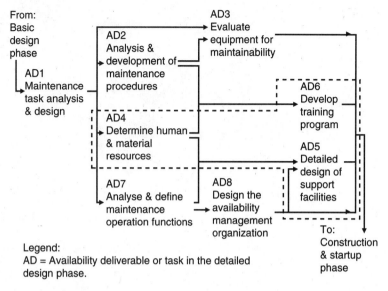

Fig. 12–1 Detailed design phase availability engineering deliverables for determining resource levels for operational availability.

Figure 12-1 shows that the deliverables associated with this requirement are as follows:

- Development and use of models to determine human, material, and other support resources levels (AD4).
- Development and use of models to determine the detail of maintenance support facilities (AD5).
- Development of a training program consistent with the decision for human resources (AD6).

AD4. Determine the Human and Material Resources for Maintenance Operations

Definition and Purpose

The maintenance task analysis determined "what" resources are required to maintain the plant. The availability engineering process must now determine the most profit-effective "amount" of human and material resources.

The resources of interest are:

- Personnel.
- Standard and special tools,
- Spare and repair parts, components, and materials
- Testing and handling equipment.
- Support facilities.
- Technical materials.

The "Opportunity" to Calculate Resources

The acquired opportunity.   Previous deliverables have left in place two basic inputs to this deliverable. They are as follows:

- Maintenance task analysis (AD1) has determined what resources are required for each maintenance task and its steps. It has collected information in a manner that allows the computation of resources.
- A knowledge of the frequency and the relationships of failures and their significance to availability has been determined by analysis and calculation during the basic design phase.

Thus, the owner has "acquired rights" to a terrific opportunity. That is the ability to actually calculate rather than make intuitive and historical ratio decisions for resources. A substantial life-cycle plant cost and use of assets can be controlled without inadvertently reducing plant availability. This can increase plant income by the expense no longer required to manage excess resources. It will concurrently increase the productivity of working assets.

Without the analyses of the basic design phase, this deliverable is not currently possible. It only becomes somewhat possible after many years of operations reveal statistical trends of use. Even then the possibility is still limited for the following reasons:

- Necessary plant experience may not have been observed and its information managed from the platform of a living availability design.
- The operating condition of most production facilities changes over their producing life. Therefore, the unplanned accumulation of historical data reflects different systems of operating conditions than the current case.

Direct and indirect financial ramifications. The ramifications of avoided asset balances and costs is both direct and indirect. An example of direct expenses is the maintenance operation payroll. Indirect expenses can be much greater. They include the functions and systems for managing resources. There is also the cost of lost business opportunities. Opportunities pass by as owner resources are absorbed in the low productivity of working assets.

A good example of missed opportunities is the plant spare/repair parts inventory. Many feel that the stocking levels for process plants are traditionally excessive. They also feel that this is by orders of magnitude.

A rule of thumb in materials management is that the annual expense for managing an inventory should be 20 to 40 percent of its value. This means that for every $10 million in excess parts held in stock, there is an annual reduction of plant and operating company income ranging from $2 to $4 million. At the same time $10 million of the owner's working assets are nonproductive. These same assets will in some way incur lost opportunity cost. There will also be additional indirect management expenses both in and outside the boundary of the plant.

Relevance of Manufacturer Recommendations

Manufacturer recommendations for material resource levels are commonly provided. However, to be realistic such recommendations must be determined in the context of the plant availability design.

The relevance of manufacturer recommendations can be improved if the procurement process provides them with appropriate information. This includes plant and equipment availability goals, criteria, parameters, and design checklists applicable to the subject item. Relevance can be further increased if the engineer's and the owner's practices include manufacturers in the development of specifications. However, even these improved recommendations should not be considered as nearly equivalent to those possible from this deliverable.

Categories of Resources and their Analyses

Basically, there are two categories of resources. They are as follows:

- Service resources include people, testing and handling equipment, tools, facilities, and technical documents.
- Consumable resources include spare/repair parts, components, and other maintenance materials.

Both are typically supplied to each maintenance task. However, there is a somewhat different approach to determining their levels. Probability distribution functions are utilized in both cases. The calculation of service resources draws upon reliability and maintainability probability distribution functions. They also require queuing rules and methods. Consumable resources also draw upon the probability distribution of reliability. However, these are distinguished by their use of inventory and control methods.

Tasks of the Deliverable for Determining Human and Material Resources

The tasks to determine human and material resources are described in the following sections. They are presented as a flowchart in Figure 12-2. The flowchart includes sequences and iterations with tasks in parallel and previous deliverables.

AD4.1. Design the scheme for modeling resources. The calculation, testing, and evaluation of resources is a substantial piece of work. However, the power of the modern computer to extract data, enable the use of models, use Monte Carlo simulation, etc., makes the task possible.

The deliverable involves industrial engineers, statisticians, and applied mathematicians. These experts form a team along with maintenance planning professionals. The team's goal is to develop a model that reflects the peculiarities of the plant and its availability scheme.

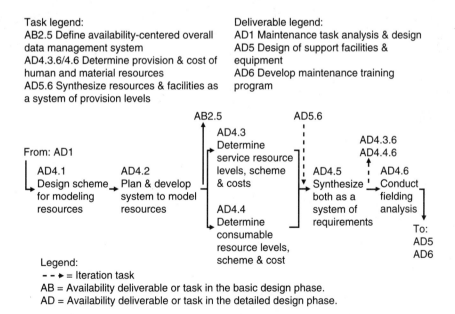

Fig. 12–2 Tasks for determining resource provision levels and costs (AD4 of Figure 12-1)

Therefore, the first task is to develop the scheme for modeling resources to define how pertinent operational research design methods will be utilized and integrated.

AD4.2. Plan and develop the system for modeling resources. This task is to define the model and its system. It is defined and developed for initial use and then as a management tool over the plant's life. The model will ultimately become part of the availability-centered data management system. Figure 12-2 shows that the results of this task flow to the one in the basic design phase for defining the overall data management system (AB2.5, Chapter 5).

AD4.3. Determine the service resources scheme, levels and costs. Figure 12-3 shows the subtasks for determining a scheme, provision levels, and costs for service resources. They are as follows:

AD4.3.1. Extract resource items from the maintenance task analysis.
The computation of service resources begins with the tabulation of resource items. These are personnel and skill levels, maintenance equipment, and technical materials. The database of the maintenance task analysis (AD1) provides the raw data. Thus, the total needs for each item are extracted from it.

AD4.3.2. Formulate the probability distributions for frequency of application. Formulate a characteristic distribution function for the fre-

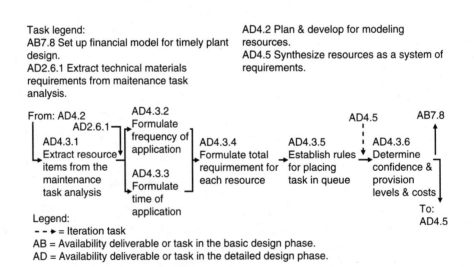

Task legend:
AB7.8 Set up financial model for timely plant design.
AD2.6.1 Extract technical materials requirements from maitenance task analysis.

AD4.2 Plan & develop for modeling resources.
AD4.5 Synthesize resources as a system of requirements.

Legend:
- - ► = Iteration task
AB = Availability deliverable or task in the basic design phase.
AD = Availability deliverable or task in the detailed design phase.

Fig. 12–3 Subtasks for computing provision levels and costs for service resources (AD4.3 of Figure 12-2).

quency of failures leading to the requirement for each resource item. The object is to determine the frequency with which maintenance requirements are expected to enter the queue line for each resource. The data is extracted electronically from the database of reliability parameters.

AD4.3.3. Formulate the probability distributions for required time of application. The probability distribution function of expected time requirements is formulated for each resource item. The result is a distribution of time that each resource is expected to be applied in each maintenance task. The input data is electronically extracted from the maintenance task analysis.

Figure 12-2 shows that detail for technical materials comes via the task of extracting the full list of requirements to be formulated as various types of technical documents (AD2.6.1, Chapter 11). This information is combined with the frequency and time-of-use processes of this and the previous task.

AD4.3.4. Formulate probability distributions of the total requirement. The task of computing the requirement for each service resource is easy to visualize. The resource analyst now has the following pieces of information:

- Probability distribution that shows how often each resource will be needed.

- Probability distribution function that shows the amount of the time that the resources will be applied to a maintenance activity.

These two distributions are combined. The result is a distribution function of the total requirement for each resource. In other words, the number of events multiplied by the time a resource is required for each event. The result is the total need for a resource.

AD4.3.5. Establish rules for placing each maintenance task in a queue. The problem of the confidence interval is a design issue that will be treated in a later task. It is partially affected by decisions for a set of queue line rules. These define where each arriving failure is placed in the waiting line for service resources.

FMECA classified and ranked failures in terms of their consequences. These will help the planner determine where to place each failure in line for service. Placement will be according to the following priorities:

- Work that is necessary because its delay may lead to a production shut-down or significant reduction.

- Work that is necessary because of the hazards they create or the damage they may cause to other equipment and assemblies.

- Work that is necessary because the plant's instantaneous characteristic availability distribution is significantly shifted or reshaped. This reduces the expected availability performance even though the plant's production level has not been reduced. This determination is made with the on-line availability model for inherent availability (A_i).

- Work that can be delayed until there is a forced shut-down or an opportunity to combine it with another maintenance task.

- Work that can be delayed until resources are available by normal acquisition processes.

- Work that can be left to a period of scheduled shut-down maintenance.

Moving an arrival forward in the waiting line for services will increase the confidence level of resource availability. There will be a much smaller percentage of time that resources will not be available. When resources are not available, the time in line will also be shorter. Meanwhile, a lower criticality item will have to wait longer in line for resources. It will also experience a longer relative wait in the line.

AD4.3.6. Determine confidence and provisioning levels and associated cost. The next requirement is to calculate the provisioning level for each resource item. This is not as simple as dividing, for example, the total hours demanded of a personnel skill category by work months or days.

Establishing the level of confidence that resources will be available is an associated design requirement. The confidence level is the percentage of time a resource is expected to be available. For example, it may be expected that 80 percent of the time there is the probability that resources will be available to repair tasks arriving for service. What if only the statistical mean of the frequency and time-to-repair distribution functions were used to compute requirements? The result would be that 50 percent of the time there will be a waiting time for resources to become available. And 50 percent of the time there will be idle resources waiting to serve.

This may be acceptable for some repairs and resources. It will not be acceptable for others. An example of the latter are failures that FMECA has found to have a high rank with respect to an important consequence.

The analysts will compute the interrelated confidence levels with respect to waiting time for the various criticality levels. Queue line rules are incorporated in this determination. As mentioned, moving an arrival forward in line increases the confidence level of its service.

Management is involved in the computation. This is because of the importance of confidence levels to overall production and business performance. The role of the resource analyst is to help management explore the strategic and financial significance of various confidence levels. It is management's responsibility to make the final decisions.

Confidence and provision-level decisions are ultimately made along with the determination of required expenses and investment in working assets. Thus, this task is to calculate expense and asset values as an integral part of its work products. Cost and asset analysis and the confidence and provision-level decisions may become iteration loops. These will occur in the process of arriving at a final decision.

- Figure 12-3 shows that the results enter the financial model (AB7.8, Chapter 7) as follows:
- Direct expenses from the levels that service personnel are staffed flow directly to the income statement subsystem of the financial model.
- Asset balances for assets such as maintenance equipment flow to the balance-sheet subsystem. Over time, they flow to income as they are depreciated.
- Working asset balances associated with service resources flow to the balance sheet subsystem.

The figure also shows that the results of this task are subject to revision by iteration. A later task will integrate service and consumable resources as a system of resources (AD4.5). The synthesis is necessary to assure that the decision is compatible across all resources. For example, two items could have inconsistent service levels. Consequently, the provisioning of one item is either excessive or a bottleneck vis-à-vis the desired availability performance.

AD4.4. Determine consumable resources scheme, levels, and costs. The approach to determining stocking levels for spare/repair parts, components, and materials is somewhat different than that for service resources. Figure 12-4 shows that the subtasks are as follows:

AD4.4.1. Extract resource items and frequency from maintenance tasks analysis. Extract the pertinent parts, components, and materials from the maintenance task analysis detail. The result will be a list of all

Task legend:
AB7.8 Set up financial model for timely plant design.
AD1.4.8 Estimate direct cost of labor & material for each maintenance task.
AD4.2 Plan & develop the system for modeling resources.
AD4.5 Synthesize resources as a system of requirements..

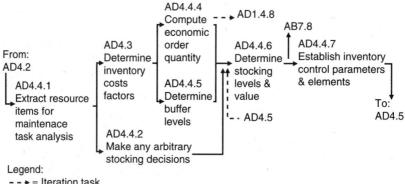

Legend:
- - ➤ = Iteration task
AB = Availability deliverable or task in the basic design phase.
AD = Availability deliverable or task in the detailed design phase.

Fig. 12–4 Subtasks for computing provision levels and costs for consumable resources (AD4.4 of Figure 12-2).

resource items and the frequency they are needed. The latter is the rate of use formed as a probability distribution.

AD4.4.2. Make any arbitrary stocking decisions.   There are arbitrary stocking decisions to be made. Some may have been defined as early as the conceptual design phase. The maintenance operation concept (AC3) included a task for establishing repair policies and rules (AC3.2).

Examples are insurance spares, components, and materials to be stocked regardless of usage rate. They are replaced when withdrawn. Another example is resource items dedicated to a specific maintenance requirement. They are only issued to that need.

Arbitrary cases are removed from the calculation of stocking levels.

AD4.4.3. Develop inventory cost factors.   The following cost factors should be developed for each resource item:

- Base cost of the subject item. This was developed in the earlier task of estimating direct material costs for maintenance task steps (AD1.4.8, Chapter 11).

- Costs that are a function of the size and number of purchase lots. Examples are quantity discounts, transportation, etc.

- Costs of handling the resource items in and out of inventory and other storage costs. The latter includes insurance, taxes, obsolescence, and spoilage. The range of such costs can be from 15 to 40 percent of the average inventory value (Tercine, 1980).

- Costs of responding to stock outages. An example is expediting costs.
- Owner's cost of capital.
- Cost of plant outages or reduced production levels.

AD4.4.4. Compute economic order quantity. An economic order quantity is computed for each resource item. The result reflects a balance of optimization between replenishment costs and holding costs.

Figure 12-3 shows that the results may iterate to the previous maintenance task analysis task of assigning material costs to maintenance task steps (AD1.4.8, Chapter 11). This may be necessary to reflect the discount on the base price, which is a function of the selected order quantity.

AD4.4.5. Determine buffer levels. Economic order quantity is the order size that will minimize the total inventory cost. However, it is still necessary to determine the buffer levels for protection against stock outages.

This is because of the risk associated with outages and their significance to the larger production and business systems. It reflects the variability of resource use and the level of service to be provided. It is also used to allow for the resource requirements of nonsignificant failures.

Like confidence levels, buffer stock levels should not be decided by resource analysts. The analyst should assist management in making decisions for acceptable stock outages.

Analysis of buffer levels will identify where creative supplier arrangements are most beneficial. These arrangements may allow management to accept a greater risk of stock outage.

AD4.4.6. Determine stocking levels and value. When the two previous tasks are completed, the stocking level may be determined. When this has been accomplished, the inventory value of each consumable resource can be computed.

In Figure 12-4, it is shown that the findings flow to the plant financial model (AB7.8, Chapter 7). They will then flow into the model's balance sheet subsystem to become a working asset.

The figure also shows that the stocking levels are subject to revision in response to the later task of synthesizing the stocking level decisions for all resources (AD4.5).

AD4.4.7. Establish inventory control parameters and elements. The following inventory control parameters and elements should be established:

- Policy for the size of replenishment orders in terms of fixed or varying orders. The fixed order is initiated when stock falls to a specified level. The varying size order is initiated at a specified interval.
- Frequency of replenishment orders in terms of taking inventory and placing orders. These are not necessarily concurrent events.
- Frequency that the stocking level and the demand are to be reviewed. In other words, when do the availability management cycles return to the provisioning analysis deliverable and tasks? These are the ones for determining the frequency of demand for parts, etc., and recomputing the level of resources (AD3) that serve that demand.
- How the review cycle will be triggered.

The last three inventory control elements are of major importance to stocking levels. Over time, the average stock levels can be dynamically optimized by the manageability of the inventory control system scheme. That is the case when the scheme allows the replenishment order and timing of the review to be varied in response to actual short-term demand cycles.

AD4.5. Synthesize resources as a system of requirements. The above described the computation of service and consumable resources. The design team must now account for the interdependency of resources. Any maintenance task can conceivably draw upon all types of resources. Thus, the probability of serving an arriving maintenance task is the aggregate mathematical consequence of the confidence or buffer levels established for each resource.

Therefore, the objective of the task is to assure that the decisions for some resources are not inconsistent with others. The cost incurred when provisioning resources with inconsistent confidence or buffer levels is two-fold. There will be nonproductive working assets and expenses incurred to manage them.

It has already been mentioned that this task will iterate with the earlier subtasks of establishing provision levels for service and consumable resources (AD4.3.6 and AD4.4.6).

Figure 12-2 also shows that this task will also iterate with a later one of similar purpose, the detailed design of support facilities and their associated equipment (AD5). Ultimately, the results of that task are synthesized with the resource optimization decisions of this task (AD5.6).

Possibility and Challenge of Support Facilities Analysis

Plant support facilities and their equipment must be designed, pro-cured, and constructed or installed. However, competent facilities design is a challenge because the nature and timing of the capital project does not allow a natural sequential approach. This is because of the lack of the fol-lowing:

- High-quality data and information for the location and frequency of tasks requiring support facilities from the maintenance task analysis (AD1.3.2 and AD1.4.5, Chapter 11).
- The resource levels to be housed for support functioning from the resource analysis process (AD4.4 and AD4.5).
- The full organizational functions defined and detailed in a later deliver-able (AD8, Chapter 13).

Therefore, necessary information and data are not fully available until the detailed design phase. This is even though the location and footage of facilities is logically part of the plant layout design in the basic design phase.

The result of poor design seems to most often be evidenced later as a shortage of space [Seminara and Parson, 1981]. This is not just a nuisance. It can extend the logistic time of some maintenance tasks. In turn, other resources levels can be excessive. They may be inconsistent with the capac-ity to move the maintenance task through the facilities and their equipment.

Functions Needing Facilities

Facilities serve the needs of both maintenance shops and administrative offices. These include the following:

- The needs of mechanical, electrical, instrumentation, and control disci-plines.
- Central storage of parts, tools, and equipment.
- Auxiliary shops such as local shops and storage in areas of high mainte-nance.
- Administrative offices and areas. These include: the needs of manage-ment and supervisors, foremen and group leaders, clerical and adminis-trative staff, specialists such as reliability and maintenance engineers, shop and field personnel, drawings and documents storage, conference and meeting areas, and habitability provisions.

AD4.6. Conduct fielding analysis. A fielding analysis is a sanity on the resource provisioning decisions. It is necessary because the nu of human and material resources available to the plant may be limited.

This is especially likely for human resources. For example, the s the relevant working population may be a constraint. Meanwhile, there be competing demands for workers. These can even be internal to the ject plant.

The possibilities and consequences of such cases are evaluated by ' ing analysis," which will assess the following:

- The impact of plant startup and subsequent commercial operatio existing resources and associated operations.
- The impact of a crisis in the new plant or facility on existing reso and operations.
- The impact on the plant or facility of a failure to acquire the n resources.
- Solutions for resolving any revealed fielding problems. Solution usually take the form of revisions to the design of maintenance c tions.

AD5. Detailed Design of Support Facilities and Equipment

Definition and Purpose

Many maintenance and most administrative activities occur in support facilities. Thus, facilities will affect both logistic and administ times with respect to time-to-maintain. They will also affect relia through the quality of the work product they support.

The purpose of this deliverable is to model and compute the fa resources. The solution includes the equipment housed by these facilit

Space Requirements

The space and layout of a shop is a function of the workload vis-à-vis the tasks associated with the work. Thus, space and layout are sensitive to the following:

- Shop staffing.
- Organizational design.
- Number and types of equipment.
- Laydown and movement areas.
- Workbench assignment and allocation.
- Storage of tools, parts, and documents.
- Associated office and administrative space.

Tasks of the Deliverable for Designing Supporting Facilities and Equipment

The analysis of facility location and space took place during the basic design phases. The deliverable for assessing plant layout for maintainability included the task for calculating requirements by approximate methods (AB11.3, Chapter 9). It is now possible to rigorously compute those requirements.

There are two aspects of analysis during detailed design phase. They are for the following requirements:

- Those that affect the logistic time of maintenance tasks. These reflect the integration of facilities and associated equipment with human and material resources. Thus, they are subsystems in a system of resources.
- Those that serve administrative functions. The determination of these functions is described in Chapter 13 with respect to organizational effectiveness.

The tasks for accomplishing the deliverable are described in the following sections. They are presented as a flowchart in Figure 12-5. The flowchart includes sequences and iterations with tasks in parallel and previous deliverables.

AD5.1. Design the scheme for modeling facilities and their equipment. The first task is to determine the operations research methods, mathematics, and practical approaches to analyzing facilities and their equipment requirements. The result is a process that is ultimately formed as a model.

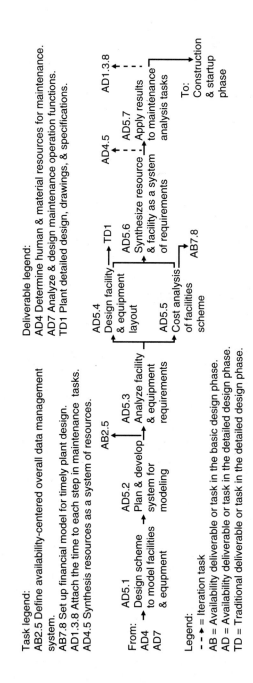

Task legend:
AB2.5 Define availability-centered overall data management system.
AB7.8 Set up financial model for timely plant design.
AD1.3.8 Attach the time to each step in maintenance tasks.
AD4.5 Synthesis resources as a system of resources.

Deliverable legend:
AD4 Determine human & material resources for maintenance.
AD7 Analyze & design maintenance operation functions.
TD1 Plant detailed design, drawings, & specifications.

Legend:
- - ➤ = Iteration task
AB = Availability deliverable or task in the basic design phase.
AD = Availability deliverable or task in the detailed design phase.
TD = Traditional deliverable or task in the detailed design phase.

Fig. 12–5 Tasks for modeling and determining support facilities and equipment (AD5 of Figure 12-1).

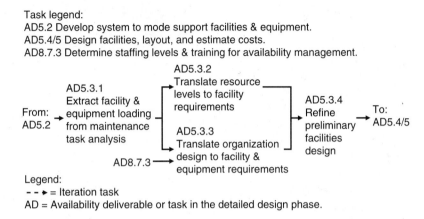

Task legend:
AD5.2 Develop system to mode support facilities & equipment.
AD5.4/5 Design facilities, layout, and estimate costs.
AD8.7.3 Determine staffing levels & training for availability management.

Legend:
- - ► = Iteration task
AD = Availability deliverable or task in the detailed design phase.

Fig. 12–6 Subtasks for the detailed design of support facilities and equipment (AD5.3 of Figure 12-5).

AD5.2. Plan and develop the system for modeling support facilities and their equipment.   The results of the previous task must be developed as a model. Thus, this task is to define the model systems to be developed for initial use and subsequent plant management. They too are incorporated in the availability-centered data management system.   Therefore, Figure 12-5 shows that the system decision will flow to the task of defining the overall data management system (AB2.5, Chapter 5).

AD5.3. Analyze facilities and equipment requirements.   The subtasks (Figure 12-6) for computing support facilities and equipment are as follows:

AD5.3.1. Extract facility-loading from the maintenance task analysis. The loading for each facility location is extracted from the maintenance task analysis. This is the statistical frequency of maintenance events. It is also the length of time the events will require the use of the assigned facility and its equipment.

AD5.3.2. Translate resource levels to space requirements. Using methods and ratios devised by the analysts, translate the computed resource provisioning levels to space requirements. These are both service and consumable resources.

AD5.3.3. Translate organizational design to space requirements. Using methods and ratios devised by the analysts, translate the results of the deliverable to design the maintenance operation functions (AD7.5, Chapter 13) and the availability management organization (AD8.7, Chapter 13) to facility requirements.

AD5.3.4. Refine preliminary facilities design. Refine the facilities scheme developed during the basic design phase. It is expected that the location of facilities is largely inflexible at this stage. It may still be possible to revise space requirements. The contents are very flexible.

The inflexibility of support facilities can be a problem. If so, it may be necessary to adjust the maintenance scheme. For example, some maintenance tasks can be assigned to other facilities outside shops, etc. The time factors for the affected maintenance tasks should be revised accordingly.

AD5.4. Design facility and equipment layouts. The facility layouts should be designed in detail. This is an industrial engineering and architecture problem. Pertinent practices include job-shop layout design and architectural programming.

Figure 12-5 shows that the results flow to the traditional project process. This is to the deliverable of preparing detailed designs, drawings, and specifications (TD1, Chapter 14).

AD5.5. Cost analysis of the facilities scheme. Any facilities scheme has its cost. It is part of the plant structure. Thus, the most relevant cost is the content. Cost is concerned with initial acquisition, continued maintenance, and eventual replacement. Figure 12-5 shows that the results flow to the financial model (AB7.8, Chapter 7) and are incorporated in the balance sheet and income statement subsystems respectively.

AD5.6. Synthesize resources and facilities as a system of resources. Human, material, and facility requirements are integrated resources. Thus, decisions for their provisioning and development are synthesized as a system of resources. Therefore, the results will iterate with the similar task for computing human, material, and other resources (AD4.5).

AD5.7. Incorporate the results in the maintenance task analysis. The maintenance task analysis included step and schedule networks for each task including all elements of the process of returning an item to service. Consequently, the results of this deliverable and the one of computing human and material resources are now incorporated in the maintenance task analysis (AD1.3.8, Chapter 11).

At this point, time has only been assigned to the active repair elements in the process of returning an item to service. This task assigns time to the logistic elements. From there, the results of facility analysis will find their way to the maintainability parameters and model.

AD6. Determine, Plan, and Develop the Training Program for Maintenance Personnel

Definition and Purpose

Maintenance personnel require training and development. Thus, training is part of service resource development. This is because a fundamental assumption of human skills provisioning is the qualifications of employees to perform.

If a training program is not carefully developed and maintained, there will be a gap between planned and actual resources. All things being equal, the plant will not be able to achieve management's specified operational availability (A_o).

Basic Approach and Concept

The training requirements are met by the following:

- Analyzing the plant maintenance tasks.
- Designing a curriculum for initial and continuous training.
- Developing training materials.
- Program implementation.
- Program evaluation.

Comprehensive training teaches both generic fundamentals and plant-specific information. Generic fundamentals cover equipment systems. They may also teach basic physical and chemical principles. Examples for a chemical process plant are heat transfer, fluid flow, and chemical properties and reactions.

Training should then move to plant-specific equipment and systems. Documentation associated with the plant process and equipment should be used. The objective is to understand the capabilities and limitations of plant equipment. It is also to inform the student why specific types of equipment and materials of construction were selected.

Mock-up models may also be part of the approach for critical maintenance tasks. They may have been initially built to evaluate plant design. Electronic models may also be incorporated in the training program.

Tasks of the Deliverable for Determining, Planning, and Developing Training Needs for Maintenance Personnel

The tasks to accomplish the deliverable are described in the following sections. They are presented as a flowchart in Figure 12-7. The flowchart includes sequences and iterations with tasks in parallel and previous deliverables.

AD6.1. Determine students and requirements from the maintenance task analysis.   The first task is to determine who will receive training and what they should receive. The maintenance task analysis included skill-level and associated training requirements for each maintenance task step. The calculation of human resource provisioning (AD4) has determined the number of people to be trained. From these, the scope and magnitude of the training program is established. This determination is for both initial and continuing needs.

AD6.2. Determine learning objectives.   Determine and group the learning objectives for initial training. The objectives are centered on the identified necessary plant skills and their certification. The knowledge required of a "new hire" is specified for each skill. Testing and evaluation materials should also be defined to assess whether a new hire has the requisite knowledge and aptitudes.

The objectives extend beyond achieving initial proficiency. They also include refresher training for maintaining proficiency.

AD6.3. Identify the instructions and procedures to be training tools. Some maintenance instructions, procedures, and manuals qualify as training tools. These are identified by this task.

Figure 12-7 shows that this may require an iteration to the task of developing instructions and procedures. Some technical materials may be subjected to additional formatting rules (AD2.5, Chapter 11). The library system and functions may also need to be revised (AD2.1) to support the training program.

AD6.4. Develop the training program.   The training program should be developed as follows:

- Establish content, program length, time structure (e.g., long sessions versus short sessions), pre-requisite training, audio-visual or computer requirements, and instruction methods.
- Develop the course materials needed to support the learning objectives. This includes instructor lesson plans, visual aids, special or modified equipment for demonstrations, student texts, simulators, and tests.
- Prepare written lesson plans. The lesson plan prevents the omission of important information and the introduction of extraneous information.

AD6.5. Develop instructor training program.   Instructors come from many sources. Some are professionals, others may require training. The primary goal is to teach preparation and classroom methods.

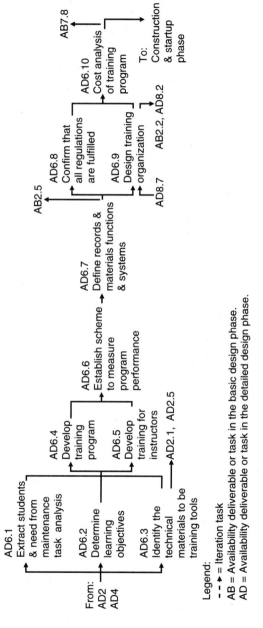

Tasks legend:
AB2.2 Develop availability element & linkage diagrams.
AB2.5 Define availability-centered data management system.
AB7.8 Set up financial model for timely plant design.
AD2.1 Define library & system for technical materials.
AD2.5 Establish rules & format for technical materials.
AD8.2 Detail processes & linkages of availability management functions.

AD8.7 Formulate the responsibilities, etc., of availability management functions.

Deliverable legend:
AD2 Needs analysis & design of technical materials.
AD4 Determine material & human resources for maintenance.

Legend:
- - ▶ = Iteration task
AB = Availability deliverable or task in the basic design phase.
AD = Availability deliverable or task in the detailed design phase.

Fig. 12–7 Tasks for developing a training program for maintenance personnel (AD6 of Figure 12-1).

It is also conceivable that professional trainers will require training. Accordingly, one objective of this task may be to assess the train-the-trainer programs of the source organizations.

AD6.6. Establish a scheme to measure program performance.   Methods and functions for measuring performance have three dimensions. They are as follows:

- To measure the performance of students both as a group and individually.
- To evaluate trainer effectiveness.
- To measure total program effectiveness.

One means for measuring effectiveness is to observe task performance in the field. Another is to identify evidence of continual improvement of performance. Both may be made part of field supervision functions. If so, it is the role of training management to audit and validate them.

Undesired outcomes in the field are also an opportunity to evaluate and improve the training program. Accordingly, functions should be established to capitalize on the findings of processes used to investigate them.

AD6.7. Define training records management functions and systems. Many people are subject to training. In turn, each is subject to substantial multiple requirements. Thus, records management functions and systems are critical to training. Poor development will undermine the resource scheme and, ultimately, plant availability performance.

Figure 12-7 shows that the task's results are incorporated in the definition of the overall plant availability-centered data management functions and systems (AB2.5, Chapter 5).

AD6.8. Confirm that regulatory requirements are served.   Some training requirements are legislated. Thus, they are important beyond just plant performance. Failure to serve them can lead to legal consequences. This can affect financial performance in the short-term. Worse, it can have long-term strategic consequences for the owner's future business opportunities.

AD6.9. Design a structure of responsibilities, roles, and authorities.   The effectiveness of the training program is partially dependent upon organizational effectiveness. Thus, this task is to define the structure of responsibilities, roles, and authorities. From there, functions and positions can be detailed.

A suitable process for this development is part of the deliverable to design the availability organization. The results of this task will ultimately be included in that organization.

Figure 12-7 shows that the product of this task flows to two others. They are as follows:

- The basic design phase task of developing a system to diagram availability elements and their linkages (AB2.2, Chapter 5). This requirement was part of the deliverable for developing the improvement, change, and data management functions and systems.
- The deliverable for designing the overall availability-management organization. Its detail has become part of the overall detailed process and linkage diagrams of availability management (AD8.2, Chapter 13). The results of the deliverable (AD8.7) iterate back to this task. The training organization is then finalized.

AD6.10. Cost analysis of the training programs.   Training is financially significant in terms of expenses and working assets. This task is to analyze the costs and asset balances generated by the training program.

Figure 12-7 shows that the findings will flow to the plant financial model (AB7.8, Chapter 7). They will enter the model's financial subsystems as follows:

- Staffing and overhead expenses will flow to the income subsystem.
- Management systems and training materials will become an asset in the balance sheet subsystem. It is conceivable that some will be expensed. Thus, they will eventually flow to the income subsystem.
- The training operation will generate a requirement for working assets associated with its staffing levels and overhead. These will flow to the balance sheet subsystem.

Bibliography

Blanchard, Benjamin S. *Logistics Engineering and Management*. 4th ed. Prentice Hall. Englewood Cliffs, N.J. 1991.

Buffa, Elwood S. and Sarin, Rakesh K. *Modern Production Operations Management*. 8th ed. John Wiley & Sons. New York. 1987.

Pack, R.W., Seminara, J.L., Shewbridge, E.G., and Gonzalez, W.R. *Human Engineering Design Guidelines for Maintainability*. General Physics Corporation. Atlanta, Ga. EPRI NP-4350. 1985.

Sanders, Mark S. and McCormick, Ernest J. *Human Factors in Engineering and Design*. 7th ed. McGraw Hill. New York. 1993.

Tersine, Richard J. *Production Operations Management Concepts*. 2nd ed. Prentice Hall. Englewood Cliffs, N.J. 1984.

13

Organization Design for Maintenance Operation Functions and Availability Management

Introduction

The previous chapter developed and applied methods for computing human, material, tools, equipment, and facility resources. They determined the position of operational availability (A_o) with respect to achievable availability (A_a). The other major determinate is organizational effectiveness.

This chapter will address the following deliverables (Figure 13-1):

- Functional analysis and design of the maintenance operation (AD7).
- Organization analysis and design for availability management (AD8).

AD7. Analysis and Design of Maintenance Operation Functions

Definition and Purpose

Ultimately maintenance operations must be carefully defined and developed. This should be done in recognition of the fundamental business purpose of such operations, which is the delivery of the plant's most profitable availability performance.

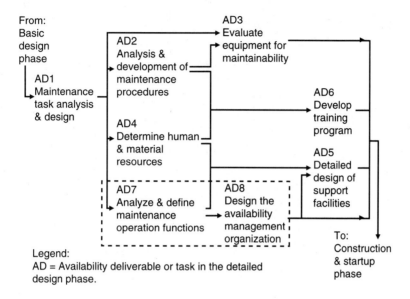

From:
Basic
design
phase

AD1
Maintenance
task analysis
& design

AD2
Analysis &
development of
maintenance
procedures

AD3
Evaluate
equipment for
maintainability

AD4
Determine human
& material
resources

AD6
Develop
training
program

AD5
Detailed
design of
support
facilities

AD7
Analyze & define
maintenance
operation functions

AD8
Design the
availability
management
organization

To:
Construction
& startup
phase

Legend:
AD = Availability deliverable or task in the detailed
design phase.

Fig. 13–1 Detailed design phase availability engineering deliverables of organization design for maintenance operations and availability management.

Indirect Functions of Maintenance Operations

There are availability management functions that are indirectly instrumental to a profitable maintenance operation. Their nature is more closely aligned to availability management. These functions are not just limited to maintenance functions, but are associated with all aspects of the availability scheme. Thus, they and their tools are the force for maintaining and advancing the effectiveness and efficiency of maintenance operations. These functions are identified in the next deliverable.

The objective of Part 6 is to show that these are also key functions of overall production and business system management. In that context, they are important with respect to classic corporate, operating company and plant-level cycles of management decisions and implementation.

Fundamental Philosophy of Organizational Effectiveness

There is a fundamental philosophy behind the objectives of this and the next deliverable. It is that the maintenance functions must be identified and then carefully structured and planned. Otherwise, the overall availability scheme will suffer.

This is likely to be the result of the following:

- Some necessary roles or functions may unknowingly be omitted. This is especially true for important roles that are not highly visible in plant maintenance operations.
- There are likely to be conflicts of interest with respect to organizational psychologies and the responsibilities they have been entrusted with.
- The full necessary flow of information and decisions will not be possible. It will instead be short-circuited, dead-ended or slowed.
- The capacity of the organization may not match the challenges that confront it. Such a mismatch can be in terms of inadequate or excessive abilities. Both are costly.

Recall that operational availability (A_o) includes administrative processes that affect the total time-to-maintain. Therefore, this and the next deliverable are important determinates of availability performance.

Job Positions and Departments

Each reference in this and the next deliverable of "activities" should not be taken as a requirement for a job position or department. Individuals and entities are typically responsible for multiple activities. This is one reason why the principles of organizational design presented in the next deliverable are so critical. The inappropriate allocation of activities to a single management or administrative entity can reduce operational availability.

Categories of Maintenance Operation Management

Maintenance operation activities can be classified as follows (Figure 13-2):

- Field management activities are:
  planning and scheduling,
  field supervision and control of maintenance, and
  engineering.
- Maintenance resources management activities are:
  human resource management,
  management of repair/spare parts, components, and materials,
  equipment and tools management.
  technical manuals management,
  facilities management, and
  engineering.

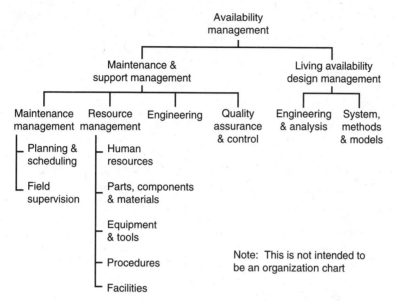

Fig. 13–2 Scope of maintenance operation functions.

- Quality assurance and control are introduced with respect to availability engineering and management. There are two dimensions to that discussion:

  How the maintenance operation functions can perform in more advanced ways.

  How availability performance and its improvement is dependent on the quality of maintenance operation functions.

Maintenance Management Functions

As mentioned, maintenance operations include the management and administrative support of field activities. These include the following functions:

- Planning and scheduling.
- Field supervision and control of maintenance.
- Engineering.

Planning and scheduling. The objectives of the planning and scheduling functions are to

Improve availability performance.

Improve the productivity of resources.

Plan for requirements before they become a crisis.

Have work performed on schedule.

The function may include a preventive maintenance manager, whose activities would be planning, organizing, and controlling scheduled maintenance (shutdown). This may also include the management of routine or routed maintenance tasks including testing, inspecting, servicing, and adjusting.

Planning and scheduling functions are responsible for three types of maintenance planning. They are daily, annual and long-range planning. Their nature is as follows:

- **Daily planning and scheduling:** Daily planning is responsible for
  - insuring that a task scope is defined to solve the problem;
  - allocating people and resources to tasks;
  - interfacing with the tag-out function of plant operations functioning;
  - developing detail prior to the maintenance task start date including instructions, resources, event dates, and the work order;
  - managing capacity and workload fluctuations; and
  - taking advantage of the opportunity to utilize temporarily excessive resources.

Planning is the beneficiary of maintenance task analysis (AD1). Its detail serves as input for the planning of current tasks.

Short-term planning is also the beneficiary of the availability model, which is used to assess the significance of each current failure. The most relevant modeled measure of on-line performance is inherent availability (A_i), which is the availability expected between scheduled shutdowns. Recall that it also assumes a perfect maintenance environment. Thus, it focuses on the significance of each new failure.

This focus is important because expected availability is reduced with each failure. However, the consequences are not always felt until the plant's production is reduced or must cease. This point was introduced in Chapter 1.

On-line inherent availability analysis allows the planner to fully understand the true importance of each new failure. This determination is made in the context of the failures that already exist in the plant as a system. This would otherwise not be possible.

Of course, the daily planning functions will draw upon the various business systems for maintenance operations. In fact, maintenance task analysis places its product in their databases.

- **Annual planning:** Annual planning is the allocation and determination of staffing and other resource levels. It is also planning for major maintenance events. This planning level draws upon the availability and financial models, the maintenance logic-tree analysis and resource calculation models.

  Modeling of achievable availability (A_a) is used to evaluate alternate approaches to major maintenance events. Chapter 9 described availability as a peaking curve. The curve is a plot of possible plant performance. Actual performance reflects the plant's location on the curve. It is a function of electing scheduled (time-based) versus unscheduled maintenance for failures. The latter is the issue for annual planning.

  Modeling of operational availability (A_o) is used for assessing and forecasting required human, material, equipment, tools, and facility resource levels. Chapter 9 also showed that planning determines where operational availability (A_o) will be positioned relative to the achievable availability (A_a) curve. The latter requires that the planned production levels be translated to their consequences for achievable availability.

  The issue of relative position is important. Chapter 9 showed that there is a ceiling to operational availability. Beyond that point, additional resources have no effect. Their expenditure only results in lost income and diminished productivity of working assets.

  Therefore, annual planning will use the availability models to avoid two setbacks:

  - operating the plant to the right or left of the availability peak, and
  - attempting to raise operational availability beyond its upper limit.

- **Long-range planning:** Long-range planning is a part of larger corporate and operating company planning functions. Both the "hard" and "flexible" elements of the plant are subject to revision.

  The consequence of the planning function can range from the realignment of the maintenance operation scheme to the partial redesign of the plant. Thus, long-range planning is part of availability management rather than just maintenance operations management.

  Therefore, the driving issue of long-range planning is the relative positions of achievable and operational availability. Thus, the capability for living availability design and analysis has the following applications:

  - Quantifying the rate that changing operating conditions over the long-term will force achievable availability downward.
  - Quantifying the rate that management will need to push operational

availability upward. This is the rate that the plant's capacity for availability performance is being progressively utilized.

- Making strategic decisions for repositioning the two curves. This will be the case as the above trends converge to an upper limit for operational availability.

The difference between these planning functions is the breadth of their impact. Daily planning utilizes existing conditions to return plant items to service. Annual planning is concerned with achieving a level of performance with respect to the plant's current capacity for availability. Long-range planning will possibly revise that capacity. It will be driven by the goal to plan the gap between possible and actual performance.

Field supervision and control.   The objectives of the functions of supervising and controlling field tasks are as follows:

- To supervise and inspect work.
- To assure that the information required at the completion of the work order is complete, correct and transmitted to the data collection system.
- To approve the finished work and assure that the associated work order is appropriately processed.
- To provide feedback on the effectiveness and efficiency of maintenance of support functions.

The functions will span all plant tasks falling in the maintenance domain. This is so even when some are assigned to the operations departments.

Field supervision functions are crucial to maintaining the integrity of the availability scheme. This is because field events determine if reality matches what is planned. This has the the following dimensions:

- Predicted reliability of plant items is partially the result of minimizing human error in maintenance. It causes plant availability performance to be distorted by the quality of the repair task. Poor work can introduce infant mortality and increase failure rates.
- Supervision is also key to assuring that time-to-maintain is kept within the planned interval.

These functions are key to meeting the specified availability performance that the engineering processes determined to be possible. This control assures the received feedback is actually a reflection of the planned system. This feedback is critical because it confirms that the assumptions in the current availability calculation are reasonable. From there, the availability scheme and its assumptions may be proven and improved.

Resource Management Functions

The second category of maintenance functions are those of resource management. They are the following:

- Human resources management.
- Management of repair/spare parts, components and materials.
- Equipment and tools management.
- Facilities management.
- Management of instruction and procedures manuals.

Human resource management. The human resource management function has immense breadth. It is not possible to explore all of its functions in this book. However, its most obvious responsibilities are hiring, firing, developing and training, promoting, administration, etc.

The roles of human resource management are dispersed throughout the plant, operating company, and corporate entities. This is because its holistic view is concerned with the following as a system of organizational issues:

- The interest of stakeholders such as employee groups, management, shareholders, government, community, etc.
- Existing situational factors such as the nature of the work force, business strategies, laws, etc.
- Human resource management policies for the flow of personnel through the organization, capacity for employee influence, reward systems, etc.
- The outcome of human resource policies with respect to employee commitment and congruence with organization direction, competence, and cost-effectiveness.
- Long-term consequences of the human resource scheme with respect to the effect on individuals and society, organizational effectiveness, etc.

Many of these issues are part of overall organizational development. Thus, they are naturally integral to the next deliverable.

One aspect of human resource management is the replacement of personnel lost due to attrition. It is also to respond to both increasing and decreasing demand for personnel. The maintenance task analysis and resource analysis deliverables are important to this role. They provide a means of forecasting personnel requirements in terms of timing, magnitude, skills, and aptitude.

Management of spare/repair parts, components, and materials. The management of spare/repair parts, components, and maintenance materials is a critical resource management function. Its functions have many facets. Some feel that they have historically been a significant source of reduced plant productivity [Herbaty, 1983]. Such consequences would be reflected in operational availability (A_o).

The functions for parts, components, and materials are as follows:

- **Provisioning:** Provisioning is the function of determining and establishing the need to stock parts, components, and materials. This function is served by maintenance task analysis (AD1).

- **Purchasing:** Purchasing is concerned with sources, pricing, etc. One goal is "smart buying." It depends upon being able to identify and plan common purchase requirements.

  One key is knowing all uses of any resource item. Having a sense of the usage rate is also crucial. The former information comes from the maintenance task analysis system. The latter is extracted from the consumable resource analysis system.

- **Expediting:** Expediting is the verification and tracking of the progress of an order until it arrives at the receiving dock.

- **Receiving:** Receiving determines that parts are received in the proper quantity and quality. This is a singular process. The receiving staff determines that parts are received and in proper quantity. The inspection staff evaluates the quality of the received items. One reason for this distinction is that inspection is often a technical or specialized process. Receiving is not.

- **Kitting:** Kitting is the function of pulling and collecting parts, etc., as a kit. This is done in preparation for an identified maintenance task. Requirements are identified by the work order.

- **Stock room management:** Management of stock rooms and bench-stocks is a function generally responsible for the jobs associated with receiving, inspection, storing, inventory control tasks, kitting, issuing, and residual materials.

- **Inventory control:** Inventory control is concerned with calculating and controlling stocking levels which are the work products of the deliverable and living systems for determining consumable resource requirements (AD4.4, Chapter 12). Inventory control will likely manage and utilize these capabilities. The function's full contribution to business results depends on its use of these living availability design subsystems.

- **Cataloging:** The parts and component cataloging function and system are a key element of parts and materials control. The primary objective is to maintain a catalogue of parts and components. This is to assure that all parts are identified according to uniform descriptions and details. This gives management the ability to analyze and search for opportunities to standardize parts; avoid treating the same part as two different parts, thus increasing stocking levels; and avoid false stock outages and consequent costs such as rush orders, subsystem outages, etc.

The detail captured and maintained in maintenance task analysis systems provides the base information on which the catalog is developed. This function involves maintaining the integrity of the catalogue and its source data.

Equipment and tools management.   The equipment and tools management function is responsible:

- Acquiring and maintaining the stocking levels of equipment and tools,
- Maintenance and calibration of testing equipment,
- Transportation and recovery of tools and equipment, and
- Staying abreast of new developments and technology with respect to tools and equipment.

This function both benefits from and manages associated segments of the maintenance task analysis and service resource analysis systems. The former identifies all applications. It also collects all background information for their management. The latter availability systems allow the function to determine necessary provisioning levels.

Facilities management.   Another resource management function is maintaining facilities. This function may be the same as the previous one. This is because the problems and concerns are the same with respect to the associated tools and equipment.

Instructions and procedures manuals management. Maintenance instructions and procedures are basic to the plant information system. Thus, their management is a crucial support function. It must be concerned with the integrity and improvement of technical manuals. It must be equally concerned with the library system, and its integrity and improvement.

The design of this function is a task of the deliverable for developing technical materials and associated management systems and organization (AD2, Chapter 11).

Engineering

Maintenance management must include engineering functions as in any other technology. They often deal with special reliability and maintainability problems. The work of the engineering functions will span daily, annual, and long-term horizons.

Consequently, the engineering functions will greatly utilize the system of living availability design as a source of tools for dealing with the planning, analysis, and design for change.

Figure 13-2 shows that the engineering function can logically appear in two places. However, the next deliverable (organization design) will show that all engineering roles should probably not reside in one or the other. This demarcation generally reflects the time horizon of assigned activities. It may also be a function of whether its tasks are generally routine or ad hoc in nature.

Quality Assurance and Control

Quality assurance functions are part of maintenance and resource management operations. They may be specific to various functions or they may be formed as a function concerned with overall plant functioning.

The concern of quality assurance functions is whether the right things are being done and if they are being done right. Thus, they are not just responsible for confirming that tasks were done correctly. They must also be concerned with whether management has tangible proof that this is normally the case.

This function is heavily incorporated in availability engineering and management. Practices documents have been developed for its use. Availability modeling and FMECA have uncovered the areas where performance is sensitive to plant practices. Thus, the quality assurance function will focuses on these areas.

Computer Systems and Integration

Many computer systems are available to serve resource management functions. They serve the needs for:

- Equipment information.
- Work order management.
- Maintenance management and planning.
- A work package library.
- Cost information.
- Materials and purchasing management.

Each incorporates parts of the previously described systems for availability engineering and management. Thus, it is necessary to identify the systems to be used. They are then integrated into the overall availability engineering and management process. Ultimately they are planned for and incorporated in the availability-centered data management system.

This determination was first addressed in the conceptual design phase. A task formulated the concept that was developed for the configuration of an overall system scheme (AC2.10, Chapter 3). It was then refined as its subsystems were rigorously defined by later deliverables.

Tasks for Analyzing and Defining Maintenance Operation Functions

The tasks for accomplishing the deliverable are described in the following sections. They are presented as a flowchart in Figure 13-3. The flowchart includes sequences and iterations with tasks in parallel and previous deliverables.

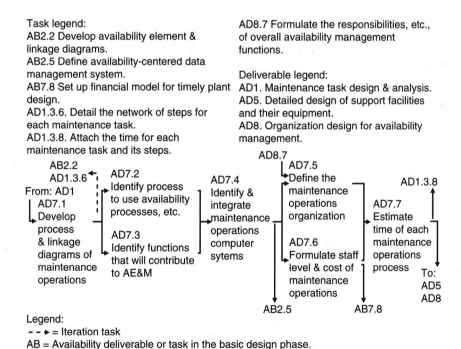

Task legend:
AB2.2 Develop availability element & linkage diagrams.
AB2.5 Define availability-centered data management system.
AB7.8 Set up financial model for timely plant design.
AD1.3.6. Detail the network of steps for each maintenance task.
AD1.3.8. Attach the time for each maintenance task and its steps.

AD8.7 Formulate the responsibilities, etc., of overall availability management functions.

Deliverable legend:
AD1. Maintenance task design & analysis.
AD5. Detailed design of support facilities and their equipment.
AD8. Organization design for availability management.

Legend:
- - ➤ = Iteration task
AB = Availability deliverable or task in the basic design phase.
AD = Availability deliverable or task in the detailed design phase.
AE&M = Availability engineering & management

Fig. 13–3 Tasks for analyzing and designing maintenance operation functions (AD7 of Figure 13-1).

AD7.1. Develop process and linkage diagrams for maintenance operations. The first task is to develop rigorous process and linkage diagrams of all maintenance operation functions. The objective is to flowchart all administrative, logistic and management activities. The results are reconciled with the maintenance task analysis. An earlier task defined the network of steps in each maintenance task (AD1.3.6, Chapter 11). The results of this deliverable are used to refine that detail.

The results of the task flow to the overall diagrams developed as part of the availability-centered improvement, change, and data management tools (AB2.2, Chapter 5).

AD7.2. Identify the processes that will utilize availability tools, processes, and systems. Many of the maintenance functions directly capitalize on the living availability design. They are short, middle, and long-term planning functions.

This task is to identify such functions. It is then to rigorously detail their use of availability tools, processes, and systems.

AD7.3. Identify roles of maintenance operations in availability engineering and management. Availability management depends on some maintenance operation processes to maintain and improve the integrity of the availability scheme. An example is field supervision.

This task is to identify such cases. Accordingly, availability management must assure that they are rigorously detailed in the process and linkage diagrams for maintenance operations.

AD7.4. Identify, select, and integrate maintenance operation computer systems. Maintenance operations make use of different computer systems. As mentioned, some of these are also systems serving availability engineering and management functions.

Therefore, this task is to make final systems and configuration decisions with respect to maintenance operations. The issue is which offered systems will be acquired and how they will be integrated.

Figure 13-3 shows that the results of this task flow to the one to define the availability-centered data management system (AB2.5, Chapter 5). The overall system scheme is refined accordingly.

AD7.5. Develop a preliminary structure of roles, responsibilities and authorities. A structure of functions, responsibilities, authorities and roles is developed for the maintenance operation. However, it is considered as preliminary. This has been the case for other deliverables in which an organization was defined. This is because the next deliverable (AD8) is to design the entire availability organization. It will include the maintenance organization as an organizational subsystem along with others. The final task of the deliverable (AD8.7) will refine the conclusions of this task.

AD7.6. Estimate staff levels and maintenance operations costs. The preceding tasks have defined the maintenance operation. This task is to estimate the cost.

Figure 13-3 shows that the results will flow to the plant financial model (AB7.8, Chapter 7). Staffing becomes part of the income statement subsystem. They also affect the balance sheet subsystem. This is the case for working assets required to support maintenance operations payroll and overhead. Assets such as computer systems also enter the balance sheet subsystems. They then flow to the income subsystem over time.

AD7.7. Estimate time for each process in the linkage diagrams. An earlier task developed detailed process and linkage diagrams for maintenance operations (AD7.1). They were then used to refine the detail of the maintenance task analysis.

This task is to assign time to each process in the diagram. Figure 13-3 shows that the detail flows to the maintenance task analysis. The objective is to attach administrative times to maintenance task steps (AD1.3.8, Chapter 11). The maintenance task analysis processes then cause these refinements to iterate back to the availability parameters and model.

AD8. Organization Design for Availability Management

Definition and Purpose

The enumeration of all necessary activities for managing overall availability performance does not automatically lead to efficient or effective organizational results. Instead, there are fundamental analytical requirements. The purpose of this deliverable is to apply them in the design of an appropriate availability organization.

Summary of Organization Design Requirements

There have been useful concepts of design developed over many years by organization development theorists. They are applied here as an analytical process concerned with the following:

- A common direction given some title as vision, mission, goals, objectives, strategies, etc.
- A detailed knowledge of work processes and linkages both within and and across the availability functions.
- Identification of the working processes in two categories of importance: those that are immediately critical to successful availability management; and those that will eventually undermine availability performance if poorly done.
- Psychological holism that will allow all important organizational mentalities to exist and balance against each other.
- The type of structure necessary to allow the flow of information and decisions between processes and organizational psychologies.
- A detailed structure of responsibilities ownership and authorities.

Three of the these concepts deserve special attention. They are the requirements for common direction, psychological holism, and flow of information and decisions. Each will be discussed in the following sections.

The tasks for organization design will subsequently be introduced. They will provide a process through which these requirements may be applied to the design.

Organizational design was identified as a task in other deliverables. The process described in the next section is applied in those cases. However, this deliverable will refine and integrate the findings of each in a monolithic scheme for availability management.

A Common Direction

Organizational design begins with setting the purpose or basic direction for availability management and its functions. The resulting statements are given many names such as mission, purpose, goals, objectives, etc. Regardless of terminology, any statement of direction must have the qualities that give individuals and small groups a sense of what is important, where to place their priorities, etc.

Figure 13-4 shows that there are minimum requirements for defining organizational direction. It also shows that they must be correlated to the

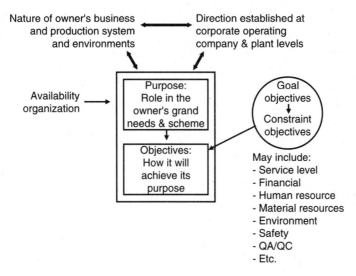

Fig. 13–4 Structure of the common direction formulated for the availability organization.

owner's larger organizational purpose and environment. The minimum requirements are defining the basic purpose of the organization, and developing a set of objectives.

Basic purpose.   The first fundamental requirement is to define the basic purpose of the organization. Although expressed in a simple statement, an organization's basic purpose is rarely immediately obvious. Consequently, the statement is just that much more critical to organizational success.

Business purpose is unique to each availability organization even though they entail the same functions producing the same work product. The reason is that the needs of each plant are different. This is because the plant is an extension of the owner's overall business strategies. Accordingly, there is a unique combination of surrounding and internal conditions facing the subject plant that affects its availability organization.

Another variable in the definition of direction is management's vision. Some owners may limit availability management to plant-level goals. Others may capitalize on the full potential of availability management. They will strive to include corporate and operating company goals in the scope of availability management.

This suggests that the purpose statement should not be just a definition of what the organization produces: availability design, management and maintenance. Instead, it should define the purpose of the organizational product with respect to the plant's needs, performance in the context of business environments, and its role in the plant owner's grand scheme.

An example is the purpose of maintenance operations. It is tempting to define it as efficient and effective maintenance of equipment. However, that is its product. Its generic purpose is to profitability deliver the necessary availability performance the plant is capable of providing. The exact nature of this purpose will vary for each plant.

Set of objectives.   The second requirement for setting a common direction is a set of objectives. These are a more definitive expression of how the availability organization will achieve its purpose.

Objectives are multidimensional. They often include financial, service level, human resource management, materials management, environmental protection, safety, and management and quality assurance objectives.

These objectives are common to every organization. The difference is their relative importance. Some are goals. Others are constraints to be recognized in achieving those goals. For example, a goal objective for maintenance operations may be associated with a service level. The remaining objectives are constraints to be worked with while still achieving the stated service level.

Conceptual phase.   Setting direction will not first occur at the time of this deliverable. One reason is that the many previous deliverables are guided by a common direction.

Direction was first addressed during the conceptual design phase. The following things occurred:

- Management established policies for availability engineering and management throughout the plant's lifetime (AC1, Chapter 3). These defined the expectations for availability management in normal plant and business functioning.
- The availability concept defined the plant in the context of the owner's larger production and business system (AC2.2). It also explored the role of plant performance in the owner's strategies (AC2.3).
- The maintenance operation concept explored and established policies for such operations in the owner's larger production system (AC3.5).

Psychological Holism in the Organization

The second organizational design requirement to be singled out for discussion is the creation of psychological holism. The goal is to allow all important mentalities and, therefore, their associated business processes to perform in good form. Thus, they will counterbalance each other in normal functioning.

Psychological holism may seem esoteric. However, weakness in this aspect of organizational design is evident in many ways. It may be one of the biggest reasons that organizations fail to thrive, achieve true quality, or achieve other goals that are crucial to its business success.

Effectiveness and efficiency.   Successful organizations are both effective and efficient. Effectiveness is the degree to which activities and outputs are correlated to the owner's markets and other realities. Efficiency is the ratio of inputs to outputs. Both are critical to survival.

The organizational elements and behavior for effectiveness and efficiency are not easily isolated and tagged. Therefore, the designer needs a framework for determining what constitutes psychological completeness for the subject availability organization and its management operations. Only then can management be assured that their enterprise will be both effective and efficient. Subsystems analysis provides such a framework especially with respect to effectiveness.

Network of roles in subsystems.   An organization is people in a network of roles. They are all related to one of five subsystems, and each has a particular significance to organizational success.

The subsystems are universal. The roles within them vary for each organization. This is because of the organization's nature and strategies. Thus, the same role in two similar organizations can be characterized by different subsystems. Furthermore, the subsystems and their roles do not rest neatly in hierarchies, functions, and departments. Nor is their influence on the organization equal or static.

The subsystem dynamic.   Each subsystem generates pressure for its own survival and advancement. This is called a dynamic. These must be understood and translated to the availability management functions. The objective is to determine how the structured relationships between organizational roles will either enable or prevent the effectiveness of the overall availability management scheme.

It is ultimately critical that each subsystem as a set of roles be structured to harness its dynamic. Otherwise, it will become a destructive force against plant and overall business performance.

The five subsystems.   Analysis of psychological holism in the organizational structure is based on the following five subsystems [Katz and Kahn, 1979] shown in Figure 13-5:

- The production subsystem is concerned with the core production process of the availability scheme. For plant availability, this is maintenance.
- The production support subsystem is concerned with the acquisition of inputs to the production subsystem, for garnering acceptance of the maintenance result; and for the external relations that aid its acceptance.
- The integrity control subsystem ties people into their defined roles. Thus, they maintain integrity of the organization's processes.
- The adaptive subsystem is concerned with the response to change.
- The managerial subsystem is concerned with the direction, adjudication, control, and integration of the other subsystems.

Method of the five subsystems.   There is a method for capitalizing on the subsystem theory. It is to define each subsystem in terms of the availability and maintenance management activities they represent. The dynamic of

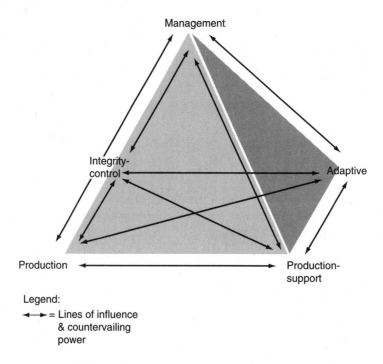

Legend:
◄─►= Lines of influence
& countervailing
power

Fig. 13–5 Five organizational subsystems and their relationship for achieving organizational holism.

the functions can then be understood and defined with respect to its subsystem nature.

This must be done for each plant. There is no universal solution to fit all availability and maintenance organizations.

A structure is finally detailed as a later step in organization design. The results of subsystem analysis will be reflected by the location of each activity in that structure. The chosen placement will help assure the efficiency and effectiveness of each. Otherwise, some will be undermined by others. This is the result of an inappropriate structuring of subsystem dynamics.

It was mentioned that any organization is surrounded by differences in internal and external cases. These include external needs, environments, and overall organization. It was also mentioned that organizational roles were not universal to each subsystem. This is because of these internal and external cases.

Therefore, it is easy to visualize the potential consequences of failing to analyze and design for psychological holism. Worse, the resulting dysfunction may be too subtle to spot as management seeks solutions to these problems. In some cases, management may not even be able to define the true symptom.

Production subsystem. The production subsystem is concerned with the core production activity of availability. Thus, the subsystem is responsible for the efficient use of resources to produce timely and effective maintenance. Its activities comprise the core productive processes of the availability organization. Therefore, it is no surprise that an availability organization may come to be identified by its main productive process. In this case, maintenance.

However, such a definition is harmful. The availability organization may come to behave strictly as a maintenance organization. Its direction may drift from achieving management's specified profit-effective availability performance.

The dynamic of the production subsystem is to maximize the proficiency of its work in terms of maintenance task accomplishment. Accordingly, it should drive toward developing standard skills and methods.

This does not mean that it will naturally rise to the highest level of potential performance. Maintenance processes may become arrested at a stage of development. This is because maintenance proficiency and the advancement of plant performance are generally conflicting goals.

Workers and functions driven by proficiency can even become negative in their attitude toward higher-level performance goals. They can come to resist stepping out of their normal roles to cooperate on matters important to total availability performance.

This too suggests the possible damage from defining the subject organization by its production subsystem. This dynamic will be able to dominate and crush other important contrary dynamics more easily. Specifically, it will diminish and destroy other roles that make it more difficult to maximize the production subsystem dynamic of proficiency.

The production support subsystem.   The production support subsystem either carries out or aids the activities to acquire resources and win acceptance of the maintenance field work product. Some of its actions are a direct extension of the production subsystem activities. This is the case for acquiring resources, planning their application, and gaining acceptance of the maintenance result. In the last case, its dynamic is to convince the remaining availability and larger owner organization that "what it gets is what it wants."

Supplier and customer resistance to current maintenance practices can also evolve as a result of the need for maintenance operations to serve their changing needs. If the pressure is successful, plant maintenance functioning will be forced to change its service level, approaches, etc.

It is important to recognize that the ultimate decisions to respond to this pressure are made in the managerial subsystem. The production support subsystem is only a secondary force for such change by virtue of generating feedback information. Otherwise, the subsystem is not a champion of change. This is because its fundamental dynamic is to sustain the field maintenance processes as they are.

The functions associated with the production support subsystem include those for planning and acquiring resources for both short-term and longer-term requirements. In this capacity, an example of its dynamic is to pressure for excess resources. Thus, the proficiency of the production subsystem is not encumbered by the need to be more sophisticated in its management practices.

The contribution of this subsystem to the production subsystem's output makes its value both tangible and visible. Therefore, it and the production subsystems are a fundamental natural threat to plant availability performance. This is because their direct and tangible nature allows them to easily dominate conflicting roles in other subsystems.

The combined dynamic will be most concerned with the proficiency of field maintenance. This is in terms of the greatest convenience to maintenance as the production subsystem. Consequently, the dynamics of the other subsystems are constraints that proficiency must be achieved within. The result is an ongoing natural conflict between contrary dynamics.

It is common to find failing or poorly performing organizations whose core production activities epitomize efficiency. This is the case when the proficiency dynamic has come to dominate the conflict.

The integrity-control subsystem. The roles of the integrity-control subsystem are directed at the organizational "equipment" for getting the work done.[1] In the availability organization, work consists of the actions of humans in accordance with established procedures. Therefore, the organizational "equipment" whose integrity must be maintained is people.

There is no guarantee that people will accept, perform, and remain in their roles. Therefore, the subsystem's roles to recruit, incorporate, motivate, develop, reward, and monitor people are part of the integrity subsystem. The result is to tie people into the overall availability management scheme as integrated functioning parts.

The production-support subsystem is concerned with providing inputs to the production subsystem. By comparison, the integrity control subsystem is concerned with whether the inputs are of an appropriate quality. This is accomplished through the selection of appropriate personnel. It then continues with personnel development, control, and reward.

Thus, the functions and elements of this subsystem are as follows:

- Human resource management.
- Procedures.
- Training programs.
- Quality assurance and reliability processes applied to field activities.

The fundamental purpose of the subsystem is to maintain integrity through the stability and predictability of the availability management organization. Therefore, its dynamic is to preserve a state of equilibrium.

The subsystem will attempt to formalize all aspects of organizational behavior. For example, when a maintenance procedure is developed and legitimized, the problems of unpredictability and instability are greatly reduced. Thus, organizational survival is insured in the short-term. Things are maintained as they are. Change is restricted.

Pressures to change often come from demands external to the availability organization. In turn, the overall availability task must change in some degree. The integrity-control subsystem is threatened by such change in the following respects:

- Some part of its activities may no longer be relevant or even required.
- Demands may be made for activities for which the subsystem is not equipped to handle.

---

[1] This subsystem would have been called the "maintenance" subsystem except for the confusion it would create in the discussion of availability management.

These threats are especially significant to the integrity-control subsystem. This is because some parts of it could easily cease to exist. And survival is a driving dynamic of all subsystems. Thus, the subsystem can easily become absorbed in its own survival. It will then strive to keep the current organizational elements that exist in place. This is even as their relevance to plant availability performance diminishes and finally becomes an obstacle.

This situation suggests the nature of dysfunction when the integrity-control subsystem is allowed to dominate. The availability organization and its maintenance operations will become increasingly rigid. They will lose their ability to respond to changing business and operating conditions, initiatives to improve plant productivity and participate in normal corporate, operating company, and plant level management cycles.

The adaptive subsystem.   The adaptive subsystem and, therefore, its dynamic is concerned with change. Thus, it is an important part of psychological holism. The other subsystems are not fundamentally concerned with response to necessary change and to ultimate destiny of the organization. They are concerned instead with availability functioning at it is.

Therefore, a subsystem must exist that is chartered to seek and identify changes in the plant's environment. These environments are:

Corporate and operating company entities that regard the plant as an asset in its overall strategies.

Sources and suppliers of human and material resources.

Other upstream, downstream, and parallel production facilities that comprise the production system of which the plant is a part. The production system of concern is both internal and external to the owner's total enterprise.

The subsystem must see both the short and especially long-term big picture. It is concerned with vision. Consequently, its functions are essentially the responsibility of top-level availability management. However, in large organizations with complex activities, they require specialists devoted to research, development, and planning. This is also the case for availability management.

Therefore, the adaptive subsystem will be characterized by the following:

- Functions associated with the entire set of analytical processes for availability engineering and management,
- The associated analysis models,
- Functions responsible for availability-centered improvement, and change, and data management systems and functions.

- The subsystem's people will manage availability engineering and management design tools, systems, and associated functions, specialists in fields such as reliability and maintenance, and long-term maintenance planning functions.

Both the integrity control and adaptive subsystems expand the basic organization. They add the specialized activities that must exist to develop, manage, and utilize the living availability design and its systems.

Therefore, both subsystems are vulnerable. Their contribution to organizational results is both short and long-term and both tangible and intangible. Their immense contribution is not generally measurable by simple cost-benefit calculations. Consequently, history shows that both are obvious targets for cost-cutting and downsizing. That is if they were ever even allowed to come in to existence. Both possibilities may be the case as early as the plant conceptual design phase.

The consequences of succumbing to this vulnerability can have a permanent effect. Other subsystems will become dominant when the adaptive subsystem is weakened or even eliminated. Once lost, this dominance will easily be able to prevent or make it very difficult to reinstate an effective adaptive psychology.

The management subsystem.   The management subsystem controls, coordinates and directs the other four subsystems. Its dynamic is to achieve their integration. It is also to adjust the total availability scheme to optimally serve its ultimate role in the owner's environment, production, and business systems. Consequently, the actions of the management subsystem do the following:

- Affect large parts of the availability organization.
- Develop rather than implement availability policies.
- Formulate rather than enforce rules.

The management subsystem will optimize the entire system. It will do so by suboptimizing and constraining the dynamics of each subsystem. Management will never be able to easily achieve perfection in this goal. This is especially so for complex organizational tasks such as availability management. Consequently, the subsystem depends upon natural inter-subsystem conflict to be effective. Accordingly, there must be an availability management structure that enables a "fair fight" between the other four subsystems. This is a fundamental objective of psycological holism in organization design.

To that end, the management subsystem includes legislative, executive, and judicial functions. It also has the task of matching plant productivity requirements with resources the overall availability organization can acquire for that purpose. Accordingly, it must set and implement policies. These requirements are founded on the findings developed by the adaptive subsystem.

The dynamic for change of the adaptive subsystem is implemented through the management subsystem. The gathering and analysis of data and the offering of recommendations are the roles of the adaptive subsystem. The decision to adopt its findings is the role of the management subsystem.

The activities that comprise the management subsystem for availability will depend on the owner's overall organization design. Its top functioning may rest with the plant manager or a position somewhere just below. However, it could just as easily rest in the operating company. The latter may be the case if the owner wishes to optimize availability performance across its plants.

Regardless of location, it is a position that will have a continuing responsibility for achieving and managing profit-effective plant availability. This is compared to having primary responsibility for a subdivision of availability such as maintenance operations.

This suggests a basic criterion for assignment. Positions that are responsible for aspects of the production, production support, or integrity control subsystems should not also be made responsible for the management subsystem. The potential for organizational dysfunction will be too great.

This criterion is not as critical with respect to the adaptive subsystem. This is because some of its aspects represent specialized roles in behalf of the management subsystem. Furthermore, if the adaptive subsystem has a marginal presence, the result of the management subsystem will also be marginal. Therefore, making such an assignment will allow the management subsystem to protect the processes that are crucial to its own effectiveness.

Application of the concept.   The application of subsystem theory is easy to visualize. The subsystems are used to identify how activities can be grouped as functions. Just as important is a determination of how they should not be grouped. The result is to assure the integrity of all activities. It is also to prevent some functions from gaining dominance.

The basic process for applying the concept is simple. It is as follows:

- Identify and understand the role of each organizational activity.
- Use the understanding of each activity to identify which of the five subsystems it is part of.

- Each subsystem has a dynamic. Therefore, define the specific dynamic of each activity. This will be guided by the nature of the subsystem it represents.
- Project how that dynamic can undermine overall availability performance generally and other functions specifically.

For example, the maintenance field operations should not house the functions that analyze and respond to changing owner needs. The result may be no true or significant response.

Another example is that the same maintenance functions should not be structured to have inappropriate influence in the determination of support resources levels. The result can easily be field maintenance supported with excessive (slack) resources.

- Finally, define how the activities must be grouped to be effective in the management of availability performance.

Structure for the Flow of Information and Systems

The discussion of this deliverable began by listing the six basic requirements or steps in organizational design. Three were to be discussed in depth before the tasks of organizational design were introduced. The following discussion is the third case. That is to determine the fundamental structure necessary to allow the flow of information and decisions between the organization's work processes.

All activities must ultimately be attached to some structure. The chosen type of structure will create varying outcomes. This is because of relative capacity of any scheme to absorb the information and then make timely, informed decisions.

The structural strategy has another implication. It determines a considerable cost of organizational functioning. The savings from an inexpensive, but inappropriate structure will be lost many times over through reduced efficiency and effectiveness. An overly capable structure will easily achieve the desired productivity, but at excessive cost. Consequently, the appropriate structure is important to the owner's financial performance.

The purpose of the following discussion is to introduce a framework of understanding on which to design the availability management organization for necessary information and decision processes. The framework is based on an information and decision processes model (Figure 13-6) for organization design developed by Jay Galbraith [1973]. It is used because it steps beyond constantly changing popular buzzwords and fads of organizational development and returns to the timeless root concepts they tend to mask.

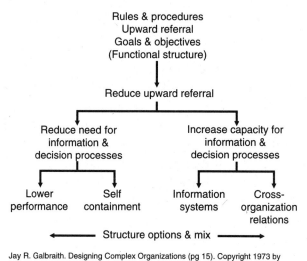

Jay R. Galbraith. Designing Complex Organizations (pg 15). Copyright 1973 by
Addison Wesley Publishing Company, Inc. Reprinted by permission of the publisher.

Fig. 13–6 Concept for information flow and decision in the organization
(Galbraith concept).

The organization design problem.   A fundamental organization design
problem is to integrate all activities in the successful execution of overall
availability management. The greater the differences between them, the
more difficult it is. This is even more the case if the activities and their
resources have some degree of uncertainty. Accordingly, more information
must be processed, more decisions made, and more actions taken. The challenge is further increased if information is also uncertain in both timing and
content.

The amount of information required for plant availability management
is a function of the following factors:

- The diversity that is inherent to tasks for availability performance. For
  example, the diversity of a field maintenance task is measured by the
  number of possible causes, challenges in troubleshooting, services
  involved in each type of failure, etc.
- The range of diverse technical specialties, capabilities, and technologies
  applicable to analysis, planning, and supporting availability management tasks.
- The difficulty level of the availability management tasks.

The degree to which these factors are pertinent has important ramifications. They are as follows:

- The number of interactions in the total availability task is greater.
- Some information will only become available to the decision processes as availability tasks progress.
- The degree that the availability organization is able to fully and exactly preplan or make advance decisions about its activities is limited.

Structures as strategies.   Different organizational forms are actually strategic variations. They are created in response to diversity, uncertainty and the need for not-yet-available information. The availability engineering design team must understand these variations and their consequences with respect to the following organizational possibilities:

- Increasing the ability to preplan the organizational tasks.
- Decreasing the level of performance required for continued viability. This may be the result of an availability policy developed by top management. It may also reflect the capacity of availability management to influence corporate, operating company, and plant-level management to accept and expect a lesser level of performance.

  Lowered performance can mean many things besides reduced quality of availability tasks. It can include excess staffing and other resources, lower acceptable plant income, etc.

- Increasing the organization's flexibility. This ability will counter the inability to completely preplan its activities. Meanwhile, it is still an election to achieve a high level of performance.

Information and decision requirements.   Organizations can be viewed as information and decision systems. This is because information and decisions are indistinguishable from the organization's system of core and subordinate work processes. Consequently, the nature of necessary information and decisions requirements for each organization are a primary determinant of what is an appropriate structure.

The relational issues of subsystem theory is critical to this design dimension. They define the necessary configuration of psychologies that information and decisions must effectively circulate through.

The management of availability involves many employees assigned to many activities distributed across many functions and multiple organizational levels. The result is division of labor and associated interdependence of individuals subjected to different influences and pressures. Meanwhile, frequent or direct communication between all affected individuals is often unfeasible.

The simplest solution would be to define each task in advance with rules and procedures. This reduces the need for communication between process participants. This is because each situation has been addressed pro-actively rather than reactively.

Unfortunately, availability management cannot completely succeed with this approach. The approach leaves no response for situations that have never appeared before or could not have been anticipated by any reasonable method and vision.

The solution is to establish additional managerial roles. Each new challenge is referred upward to a shared manager. The weakness is that each such manager is limited in terms of the information processing and decision-making he or she can effectively address.

Upward referral is much more appropriate when management progressively chooses business strategies and performance targets that are more challenging. The capacity of shared managers is quickly exhausted. With upward referral the organization will be able to respond as necessary to its chosen vision.

The next level of possible organizational strategy is to lower the decision process to the points where the work processes reside. Decisions now occur where the related information has its source. This, in turn, requires that people at these levels be additionally empowered to match this new situation.

Now there is a control problem. Management must somehow assure itself that these empowered individuals will act consistently with the overall availability mission and scheme. In other words, it must be comfortable that individuals will step beyond their focus on the specific task and instead consider the greater intention of that task.

A solution is to use goals and objectives along with rules, procedures, and an appropriate reliance on upward referral. However, the ability of individuals to function successfully within the established goals and objectives can be a challenge. This challenge reflects the degree of uncertainty inherent to the work processes and their associated goals and objectives.

If uncertainty is great, goals and objectives will have to be set and reset throughout some work processes. This requires additional decision making. The structure of upward managers must again be called upon to address new challenges. And just as before, the managers will eventually become exhausted by the demands to process information and make associated decisions.

Structural choices.   The functional challenge of availability management and its maintenance operation is generally described by the characteristics outlined in the previous section. Consequently, the availability organization design team has structural choices. The first choice is to accept the above limitations and operate as a functional organization. Alternately, it can counter these limitations by adopting one or a mix of four structural strategies.

This, in turn, leads to two fundamental choices. The relation of the choices is shown in Figure 13-6. The figure shows the following set of choices:

Reducing the need for information and decision processes. In other words, simplify the challenge to one that can be more easily served. Its two structural options are lowered specification of acceptable performance, and organizational self-containment.

Increasing the organization's capacity for information processing and decision-making. This is accepting the full challenge facing the availability organization and creating a structure that can serve its scope. Its two structural options are information systems, and work process relations across the organization.

Strategies to simplify the challenge. The first of the two fundamental choices is to reduce the information and decision processes that are needed to accomplish the organization's task. The options are as follows:

Determine opportunities to reduce performance to an acceptable lower level. Performance is reduced in terms of the outputs, inputs, and performance parameters of a workprocess.

For example, the reduction can be applicable to resource utilization. The availability organization can respond to its challenge by increasing the human, material and other support resources. This may be 6preferable to better managing a lower level of resources while still achieving the same performance.

Create self-contained organizational units. The second strategy for reducing the amount of necessary information and decisions is to organize in self-contained work processes. The strategy reduces the diversity of input, output and throughput diversity.

As the point of decision making is moved closer to the action, a difficult problem is made simpler. It is, therefore, much more manageable.

One source of simplification is that the unit is simply unaware of the full problem at hand. This is because of the expertise that remains outside the boundaries of its members.

As management's choices between business strategies make the overall organizational task more complex, the job of higher management also becomes more difficult. This is because its fundamental responsibility is to integrate the organization's self-contained units. The eventual result is the overload of these managers. This, in turn, leads to a natural limit on performance.

Strategies for increasing the capacity for information and decision-making.   The alternate fundamental strategic choice is to increase the capacity for information and decision processes. This is a chance to rise and meet the inherent challenge facing the organization. The options are as follows:

Develop information functions and systems. This strategy increases the capacity of existing information and decision processes, creates new processes, and introduces new approaches. It also increases the capacity to use information that evolves as the work process progresses.

There are two dimensions to this strategy:

- Information is collected at its source. It is then directly accessed at the appropriate time by those who have a need for it. These people may be in higher, lateral and lower organizational levels. The alternate approach is to direct information upward to a common manager and then downward to the user.

- The decision-making capacity of some entities may have to be increased to process the increased information. Analyses models serve this purpose.

Use cross-organization information sharing and decision-making. This strategy includes direct lateral contact, liaison, task forces, teams, integrating roles, managerial linking, matrix forms, and systems organization. Thus, cross-organization methods range from the simple, obvious and inexpensive to those that are sophisticated and costly. However, more complex, expensive forms are an addition rather than a substitute for the less demanding methods. This highlights the need to avoid "excessively" capable organization structures.

Using these cross-organization methods, the organization's capacity to process information and make decisions is increased. This is because empowerment is increased at lower levels and between all of the functions relevant to the work process. Consequently, cross-organization participation results in holistic rather than micro decision making.

Cross-organization forms can be contrasted with self-containment. Both increase discretion at lower levels. However, the self-contained strategy is possible when there is little necessary sharing of resources and expertise across outside functions and work processes. However, cross-organization relations still allow empowerment but without the loss of interdependency.

The four strategies compared.   There is a common denominator to all four strategies. It is to reduce the magnitude of necessary upward referral. This is done in two diametrically different ways.

The first two strategies reduce the diversity and task difficulty confronting the organization. In contrast, the third and fourth strategies accept the difficulty as a given. They instead create the requisite capacity for information and decision processes to confront and deal with the chaos inherent to the accepted challenge.

The first strategy—lowered performance—is the default. It evolves naturally if management does not consciously choose one or a mix of the four.

Many manufacturing plants are victims of the default decision. This can begin in its design phase if availability engineering is omitted from the project scope. For example, lowered performance may be because the plant is overdesigned. The same performance would have been possible with hard design and maintenance operations developed in an integrated availability scheme. Consequently, this omission is a choice by default to accept less profitable performance.

Alternately, simplification by omission may cause the plant to fall short of its specified availability performance. Another possibility is that the plant may not be able to adjust its operational availability to match changing markets and operating conditions. This is also to accept lesser results by default.

Application of the concept.   The objective of the structural strategy concept is to assess the availability management processes and linkages with respect to information and decision flow. It is then to determine the necessary unique mixture and configuration of the strategies.

In some cases, it will be possible to accept lower performance. It will be acceptable for other functions or work processes to be structured as self-contained units. The information system strategy is available and feasible through the availability-centered data management system (AB3). It will be an integral part of the other three strategies.

Finally, the strategy for cross-organization relations is studied carefully prior to its application. This is because of its natural costliness. Therefore, this strategy is the last resort. Popular literature and its philosophies would suggest otherwise. However, wholesale application should never be advised without careful determination.

Tasks of the Deliverable for Designing the Availability Management Organization

The tasks for designing the availability organization are described in the following sections. They are presented as a flowchart in Figure 13-7. The flowchart includes sequences and iterations with tasks in parallel and previous deliverablesThe above discussion identified multiple requirements or design principles. Three were discussed in some depth. The following tasks will apply all of them in a design process. The final work product is a detailed organization structure.

AD8.1. Establish a common direction for the availability organization.   A common direction is established. It is an expression of management's goals, objectives, strategies, vision, etc. This topic was discussed in a previous section. The requirement was first addressed in the conceptual design phase.

This task is probably the most difficult. One reason is that few people fully understand the principles and concepts of direction. Even when they do, the associated brainstorming and insights are always a challenge.

The work can be helped along by including strategists in the team. Ideally, these will be the same people who work in the owner's corporate and operating company strategy divisions. Therefore, they will also contribute and continue the basic strategic vision that originally lead to the project..

AD8.2. Detail the work processes and linkages of availability management.   The principles of total quality management advise the practice of diagramming the organization processes and their linkages to each other. This is an extremely effective practice. It is also a crucial step in the design of an effective and efficient availability organization.

The benefit goes beyond the detail it provides. The organization's design team is driven to initially think of the organizational task instead of nonintegrating concepts such as subdivisioning, departments, and functions.

This is an important distinction. Organizations with complex challenges require many of their functions to work laterally as opposed to within the hierarchy of departmental boundaries.

Figure 13-7 shows that input to this task initially comes from earlier tasks in other deliverables. The input defines the following organizational subsystems:

- Technical materials operation (AD2.3, Chapter 11).
- Training operations (AD6.9, Chapter 12).
- Maintenance operations (AD7.1) of the previous deliverable.

Task legend:
AB2.2 Develop availability element & linkage diagrams.
AB2.5 Define availability-centered data management system.
AB2.6 Analysis & preparation of technical materials.
AB7.8 Set up financial model for timely plant design.

AD2.3 Define organization for technical materials management.
AD6.9 Define organization for taining program.
AD7.5 Define preliminary organization for maintenance operation.

Deliverable legend:
AD7 Analysis & design of maintenance operation.

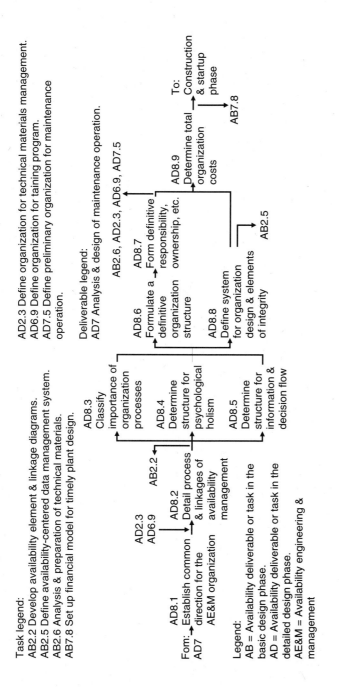

Legend:
AB = Availability deliverable or task in the basic design phase.
AD = Availability deliverable or task in the detailed design phase.
AE&M = Availability engineering & management

Fig. 13–7 Tasks for designing the availability organization (AD8 of Figure 13-1).

Process and linkage diagrams are part of the change, improvement, and data management systems (AB2.2, Chapter 5). Preliminary diagrams were developed to the extent necessary and possible in the basic design phase. Thus, they provide initial input to this and the other defining tasks. Accordingly, Figure 13-7 shows that results of this task flow back to become the final diagrams in the change, improvement, and data systems (AB2.2, Chapter 5).

AD8.3. Categorize the importance of the organization's processes. Organizational expert Peter Drucker recommends that the load bearing or key activities of the organization be identified. Therefore, the detailed processes are classified as critical, major, and minor.

This distinction is made by the following questions:

- In what processes is excellence immediately and persistently required to achieve the availability organization's goals?
- In what activities or processes would the lack of efficient and effective performance ultimately undermine the capacity to achieve the goals?

Organization design focuses on the identified critical and major activities. Focus is crucial since there are always limited resources and time. The processes and linkages diagrams are used as the "menu" from which to identify them.

AD8.4. Determine the structure requirements for psychological holism. Analysis of psychological holism is required to assure that all important organizational mentalities exist and balance against each other. Thus, the analysis identifies how the flowchart processes (AD8.2) must be grouped (not grouped) to assure their integrity. It also assures that some do not assume inappropriate dominance.

AD8.5. Determine the structural strategies for the flow of information and decisions. This task is to assess the process and linkage diagrams with respect to information and decision flow. From there it is determined how the four structural strategies are most appropriate in an overall configuration. Examples are as follows:

- It may be possible to accept lower performance in some acceptable form for functions that are classified as minor.
- Other work processes will have a nature that allows them to be placed in self-contained functions.
- The vertical information strategy is made possible by the availability-centered data management system. Thus, the designers will determine which functions depend on its data and analyses elements.

- The remaining functions will be served by lateral relations. Consequently, the goal is now to determine which of the organizational methods for this strategy will be applied to relate the subject functions and work processes. The methods include direct lateral contact, liaison, task forces, teams, integrating roles, managerial linking, matrix forms, and organizational systems. The last method is to temporarily and permanently combine individuals, functions, organizational units and external organizations as a system. It is a dynamic form. Membership is determined by current requirements. Participation of each entity can be short or long-term.

This and the previous two tasks are guided by the findings of the task of establishing a common direction for the availability organization. This guidance enables the design team to make the necessary design distinction and decisions.

AD8.6. Formulate a definitive organization structure.   The previous two tasks are not yet a working diagram. They must now be converted to a classic organization chart. However, it is important to note that only now is such an instrument formulated. There is always a temptation to leap to this step. However, such a leap will most likely build into or leave fatal flaws in the final design.

AD8.7 Formulate the detail of responsibility, ownership, and authority. Organization design will finally overlay the detail of responsibilities, ownership, and authorities on the organization chart. The subtasks are as follows (Figure 13-8):

AD8.7.1. Prepare a customer, supplier, and ownership matrix. In a matrix, with the business processes as a dimension, identify the organizational participants, responsibilities, authority to make decisions, rights to veto, and ownership.

AD8.7.2. Develop department and position descriptions. Utilize the matrix to develop departmental and position job descriptions.

AD8.7.3. Determine staffing levels and training requirements. The job descriptions are extended to identify the necessary range of staffing levels and their training requirements. These details will flow to the determination of support facilities. Another task will translate the organizational detail to the facilities required to house them (AD5.3.3, Chapter 12).

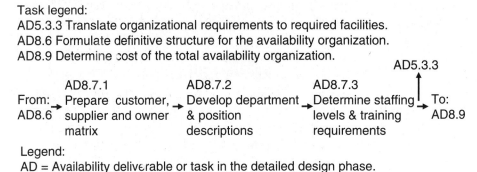

Task legend:
AD5.3.3 Translate organizational requirements to required facilities.
AD8.6 Formulate definitive structure for the availability organization.
AD8.9 Determine cost of the total availability organization.

Legend:
AD = Availability deliverable or task in the detailed design phase.

Fig. 13–8 Subtasks for formulating responsibility, ownership, and authority (AD8.7 of Figure 13-7).

Figure 13-7 shows that the results of this deliverable will iterate to the various suborganization's availability management. These were formed as a preliminary scheme for their respective domains. They are as follows:

- Improvement, change, and data management functions (AB2.6, Chapter 5). These organizations will manage the processes associated with availability parameters, failures, modeling, and optimization. Modeling includes availability, financial, resources, and facilities.
- Maintenance instructions, procedures, and manuals functions (AD2.3, Chapter 11).
- Training organization (AD6.10, Chapter 12).
- Organization for maintenance operations (AD7.5, Chapter 13).

AD8.8. Define the information system for organization design and its elements of integrity. As suggested by the previous tasks, organization design is a substantial piece of work. The design may also be dynamic over time or the organization's environment and derived performance changes. Therefore, it is necessary to have the ability to trace through its design. This is necessary when staff levels change; when there is a need to revise the structure; and when management changes some basic approach. An example of the last possibility is if management were to begin contracting for availability management and maintenance functions that were previously provided in-house.

Thus, an information system should be defined for that purpose. It should capture the design process and decisions and their various organization charts, job descriptions, etc.

Figure 13-7 shows that the ultimate system scheme flows to the task of defining the data and information system (AB2.5, Chapter 5). This is part of the deliverable of developing the change, improvement, and data management functions and systems.

AD8.9. Determine the cost of the total availability organization. Organizations have costs. Differing schemes have widely different cost structures. Thus, this task is concerned with the cost of organizational power. The more power, the greater cost. The analysis may cause the organization design team to try alternate schemes. For example, plant resource levels can be increased to offset less organizational power and, therefore, cost.

The analysis results will flow to the plant financial model (AB7.8, Chapter 7) as follows:

- The income statement subsystem will reflect staff levels and training.
- The balance sheet subsystem will reflect the working assets required to support staffing.
- The investment in electronic systems for maintaining integrity will flow to the balance sheet subsystem.

Bibliography

Beer, M., Spector, B., Lawrence, P.R., Mills, D. Quinn, and Walton, R.E. *Human Resource Management: A General Manager's Perspective.* Free Press of Macmillan, Inc. New York. 1985.

Drucker, Peter, F. *Management: Tasks, Responsibilities, Practices.* Harper and Row. New York. 1993

Galbraith, Jay. *Competing With Flexible Lateral Organizations.* Addison-Wesley. 2nd ed. Reading, Mass. 1993.

Galbraith, Jay. *Designing Complex Organizations.* Addison-Wesley. Reading, Mass. 1973.

Herbaty, Frank. *Handbook of Maintenance Management: Cost Effective Practices.* 2nd ed. Noyes Publications. Park Ridge, N.J. 1990.

Katz, Daniel and Kahn, Robert, L. *The Social Psychology of Organizations.* 2nd Edition. John Wiley & Sons. New York. 1978.

Patton, Joseph D. *Maintainability and Maintenance Managment.* 2nd ed. Instrument Society of America. Research Triangle Park, N.C. 1988

Richards, Max D. *Organizational Goal Structures.* West Publishing Company. St. Paul Minn. 1978.

Stebbing, Lionel. *Quality Assurance, The Route to Efficiency and Competitiveness.* Ellis Horwood Limited. Chichester, West Sussex, England. 1993.

Tersine, Richard J. *Production Operations Management Concepts.* 2nd ed. Prentice Hall. Englewood Cliffs, N.J. 1984.

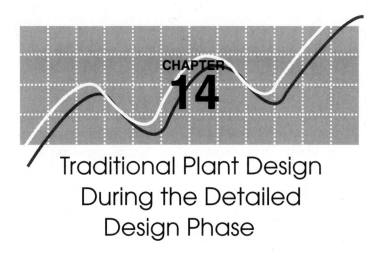

14

Traditional Plant Design During the Detailed Design Phase

Introduction

The traditional design deliverables have been presented for the conceptual and basic design phases (Chapters 3 and 10). This chapter introduces those for the detailed design phase.

However, like the previous cases, it is not the objective of this chapter to describe the deliverables with the same rigor applied to those for availability. Instead, the purpose is to acquaint the reader with the traditional capital project deliverables of the phase. It is also to identify the fundamental bilateral relationships between availability and traditional deliverables.

The intertask relationships are extensive. It is outside the scope of this book to explore them.

The Scope of the Traditional Detailed Design Phase

The traditional phase activities will design the plant in detail. The detail is packaged as construction documents. Therefore, the phase begins when the design team takes possession of the previous design phase deliverables.

They should have already been reviewed and approved by the owner. However, the basic design is still subject to minor change during this phase.

The traditional scope of the phase is as follows:

- A detailed design of all plant subsystems is completed.
- Specifications are finalized and the equipment procurement initiated for items that were not part of earlier activities for long-lead-time equipment. Activities include the coordination and approval of manufacturer documents, field expediting and quality control procedures.
- Operating and maintenance procedures are acquired or developed.
- Hazard and safety analysis is subjected to more rigorous approaches.
- Construction drawings and documents are finished, approved, and issued.
- Procurement of bulk materials may be initiated.

The traditional capital project deliverables are shown in Figure 14-1. The deliverables are coded as "TD" to denote a "traditional" deliverable during the "detailed" design phase. Each deliverable is described in the following sections.

TD1. Detailed Design, Drawings, and Specifications

Traditional Scope

The efforts of the basic design phase and the details evolving from the final procurement decisions are converted to construction drawings and specifications. The design activities of multiple disciplines are integrated. These include piping, electrical, instrumentation, various mechanical and equipment specialties, safety and loss control, materials, and structural and civil engineering functions.

The basic tasks for developing detailed construction drawings and specifications are as follows:

- Complete the mechanical flowsheets for review and construction. The final flowsheets will also be used as a tool for updating the equipment list and instrument index.
- Finalize specifications and issue them to procurement for equipment not previously procured.
- Review and finalize the detailed plant plot and equipment layout.
- For some plants such as a chemical production process facility, update the line list and complete the detailed piping design.

Task legend:
AB1.7 Development decisions for practices documents.
AB5.1 Allocaite availability parameters to plant logic diagrams.
AB8.7 Rank the consequences of each failure in the plant.
AB10.7 Sensitivity & analysis of availability.
AD1.3.6 Detail network of steps for each maintenance task.
AD2.8.6 Release technical materials to unrestricted use.
AD3.3 Evaluate equipment for maintainabillity from human factors.
AD3.5 Assess human reliabililty in maintenance tasks.
AD5.4 Design facility & equipment layouts.

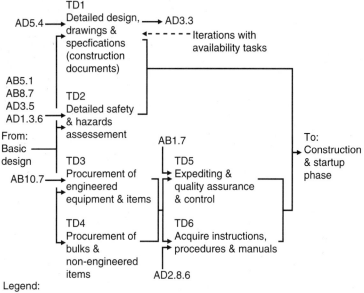

Legend:
- - ► = Iteration task
AB = Availability deliverable or task in the basic design phase.
AD = Availability deliverable or task in the detailed design phase.
TD = Traditional deliverable or task in the detailed design phase.

Fig. 14–1 Traditional deliverables for the detailed design phase.

- Prepare final utility loads and level determinations, utility flowsheets, and detailed utility system design, drawings, and specifications.
- Prepare the final design of the instrumentation and control system.
- Complete the loss control design with respect to hardware, procedures, protection of materials, and other requirements to protect against, abate, or respond to each identified hazard. The design of the plant safety warning and response system and its subsystems are part of this deliverable.
- Complete civil design for foundations, structures, grading, and drainage.
- Review the above designs with the owner, make necessary revisions, and issue them for procurement and construction.

Pertinence of Availability Design

Availability design would enable these traditional activities to give both increased and methodical attention to reliability and maintainability issues. One case is the availability task concerned with the design of equipment for human factors (AD3.3, Chapter 11).

At this stage in the project, focus is skewed toward adjusting maintenance tasks and operations to fit with human factors. This is because plant hard design has become generally solidified. In turn, some of the reliability and maintainability factors are also fixed.

However, other factors of reliability and maintainability are still flexible. They are the ones associated with maintenance operations. Accordingly, final detail would be assessed against maintenance tasks analysis and associated procedures documents. These deliverables are reviewed against the drawings and the models produced by traditional design. Equipment may also be available for such reviews at the manufacturer's facility. The maintenance tasks are adjusted accordingly.

The availability engineering process would also test and refine the layout and sizing of maintenance support facilities and their equipment (AD5.4, Chapter 12). There may be a need to reconcile the plant layout with the detailed maintenance facility analysis. The plant availability performance or scheme is subject to adjustment if plant layout is fixed.

The bilateral relationship between traditional and availability engineering has a general nature: to review the plant detail. It then refines the operational availability (A_o) to achieve management's specified performance. Any of the many availability deliverables and tasks may be affected by iterations.

The deliverable has a general connection to the availability engineering deliverables and tasks. It may affect and be affected by many of the availability tasks. It is not in the scope of this book to explore that detail.

TD2. Detailed Safety and Hazards Analysis

Traditional Scope

Detailed safety and hazards analysis begins at some later stage in the detailed design phase. Ideally, it would occur when detailed design is finished. However, schedule constraints preclude such a luxury.

Analysis activities of the basic design phase identified and assessed hazards. This was preliminary in nature. The design has now advanced considerably in its hardware, layout, and procedural detail. Consequently, the following methods are most pertinent:

- A hazards and operability (HazOp) study is useful for investigating how the plant may deviate from its design intent and create hazards in the plant. It could also search for operating and maintenance problems that would compromise the plant's ability to achieve its intended productivity.
- A fault-tree analysis can focus on investigating particular accident events. Accordingly, its objective is to determine causes of each subject accident as a function of both equipment failures and human error.

Pertinence of Availability Design

Detailed hazards analysis can be qualitative, quantitative, or both. Availability engineering would make the options for quantitative approaches readily possible because it naturally creates necessary data and information.

One input is the formulated reliability parameters that are allocated to the many plant elements (AB5.1, Chapter 6). They are important to establishing the probable frequency of safety-critical failures. Meanwhile, FMECA will reveal the chain and consequences of initiating or contributing events (AB8.7, Chapter 8).

It was mentioned that the HazOp process could search for maintenance problems that may compromise plant productivity. This expectation is not realistic. An appropriate grasp of pitfalls from maintenance cannot be well treated by a hazards and operability study without being preceded by availability engineering. To persist in this belief and the false sense of security it creates may be a source of risk.

However, availability engineering enables the hazards and operability study to fulfill this promise. This is because the detailed maintenance tasks (AD1.3.6, Chapter 11) and associated technical documents (AD3.5, Chapter 11) would be subjected to the hazards and operability study. The objective is to examine the consequence of a worker failing to properly execute each step of a subject procedure. Therefore, the hazards studies would draw upon a body of detail that would not otherwise be available.

TD3. Procure Engineered Items and Equipment

Traditional Scope

Engineered equipment is defined as designed and made to order by the manufacturer rather than purchased off the shelf. Therefore, plant layout design and detailed engineering and procurement of engineered items and equipment are potentially iterative activities.

The basic procurement activities for engineered items are as follows:

- Prepare and tender specifications. This activity is concerned with the selection of manufacturers or vendors, the preparation of specifications, and the preparation and transmittal of the request for quotation to those suppliers.
- Compare quotes and select a supplier. Quotes are received and evaluated for commercial terms and technical merit. The vendor or manufacturer is selected; detailed commercial terms are established; and delivery dates for drawings, supplier data, and equipment are established.
- Manufacturer drawings are received and routed for review by the pertinent engineering disciplines. The annotated drawings are returned to the manufacturer. Revised drawings are received from the supplier and become certified project drawings.

Pertinence of Availability Design

The traditional procurement process would be affected by availability engineering. As mentioned earlier, the purchase specifications will include explicit requirements for reliability and maintainability. The requirements may also include those for providing detail for maintenance operations. They may also specify participation in some availability design deliverables and tasks.

Availability design would also affect the selection of candidate manufacturers, subsequent choice of a manufacturer, quality assurance and control requirements, and processes to review supplier documents. For example, availability modeling during basic design may influence the supplier selection. Sensitivity analysis (AB10.7, Chapter 9) may indicate the need to stress comparative qualifications and experience with respect to assemblies and components within the subject equipment item.

It is also possible that nontraditional designer-supplier relationship programs will be made more relevant to plant design. This is because the manufacturer's expertise may be crucial input to the design and sizing of equipment for reliability and maintainability.

TD4. Procure Bulk Materials and Nonengineered Items

The procurement of nonengineered items deals with bulk materials as well as off-the-shelf items. This activity may occur during detailed design as necessary to be available for construction and startup.

TD5. Expediting and Quality Assurance and Control

Traditional Scope

Expediting and quality assurance and control are often separate functions. This is because of a potential conflict of interest. The roles of each are as follows:

- Expediting follows up on drawing submittals, parts lists, and data delivery to the document control process; possibly coordinates quality assurance and control activities; and visits the manufacturer's shop for various reasons, such as to confirm progress.
- Quality assurance and control monitors planned tests, fabrication methods, etc. This occurs at the manufacturing and construction sites.

Pertinence of Availability Design

The availability engineering process would establish additional quality assurance requirements. For example, there may be requirements for programs to test and evaluate maintenance procedures as equipment takes form at the manufacturer's facility. Of course, the full normal assurance requirements of the availability discipline would be supported by traditional assurance functions.

The availability-centered issues would have been formed by the deliverable to develop availability-centered practices documents (AB1). Ultimately, availability management will make decisions for which practices are to be developed (AB1.7, Chapter 5). They will be based on importance and urgency. Other availability tasks will be involved in the determination of importance.

TD6. Needs Analysis and Design of Instructions, Procedures, and Manuals

Traditional Scope

There are three basic approaches to production and hazards control: design, procedures and management. Since the design has progressed, the design team is now left with mostly procedural and management solutions.

Traditionally, the development of operating and maintenance instructions, procedures, and manuals may begin when the detailed design activity is approximately 60 to 70 percent complete. Experience has been that operating and especially maintenance procedures are not given the attention they deserve. New safety management regulations and industry-wide initiatives may change that condition.

Pertinence of Availability Design

One of the benefits of availability design is to insure that attention to technical materials for maintenance is appropriate and thorough. The final results of this availability deliverable are instructions, procedures, and manuals issued for unrestricted use (AD2.8.6, Chapter 11).

Summary

This chapter ends the discussion of availability engineering and management in the detailed design phase. At this point the following has been accomplished:

- An optimal availability scheme has been developed which reflects strategic decisions for the short and long-term associated positions of achievable (A_a) and operational (A_o) availability.
- The maintenance operations and resource levels associated with the optimized scheme have been determined. The business of maintenance operations is to most profitably deliver the specified operational availability (A_o) performance. Thus, the detailed design phase has developed the scheme to fulfill that business purpose.
- Systems have been developed to define and study the full life cycle cost of plant availability performance. These will be used repeatedly throughout the plant's producing life. The goal is to revisit and realign the results as part of normal operations and planning cycles. Availability management as part of corporate, operating company, and plant management is the subject of Part Six.

Bibliography

Center for Chemical Process Safety. Guidelines for Hazard Evaluation Procedures. Institute of Chemical Engineers. New York. 1989.

Center for Chemical Process Safety. Guidelines for Technical Management of Chemical Process Safety. American Institute of Chemical Engineers. New York. 1989.

Rase, Howard F. and Barrow, M.H. *Project Engineering of Process Plants.* John Wiley & Sons. New York. 1957.

PART FIVE

The Construction and Startup Phase

CHAPTER 15

Construction and Planning
for Startup

Introduction

This chapter explores the purpose, role, and requirements of availability engineering and management during plant construction and startup.

Two Cases of Construction and Startup

There are two types of construction and startup. The first is the construction or major expansion of a plant. This part of the book is concerned with this type.

The second type is construction that occurs periodically during the plant's producing life. This type of construction and startup is capital projects triggered by cycles of strategic planning, applied research and development, planning to match production capacity to longer-term demand, productivity testing and evaluation, etc. These cycles are part of managing the owner's total production system. They will be explored in Part Seven.

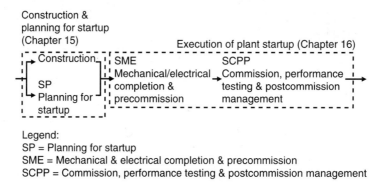

Fig. 15–1 Stages of the construction and startup phase.

Approach to Describing Availability Engineering and Management During Construction and Startup

The previous three parts of this book presented availability engineering as a network of uniquely identifiable deliverables. However, availability engineering and management requirements become hidden within the activities of construction and startup.

This part presents the network of top-level deliverables or activities for plant startup. The concerns and requirements of availability engineering and management are then identified for each. Finally, the flowcharts for traditional startup activities are reformed. The result is flowcharts of availability and management during startup (Figure 15-3, 16-3 and 16-5).

The section is not offered as a significant discussion of the construction process. This is because availability-centered activities during construction tend to be associated with startup. The subject of this book is the availability discipline. Thus, there would be no justification for an in-depth exploration of the construction process.

Stages and Goals of Construction and Startup

Figure 15-1 shows the construction phase along with the startup stages. The diagramed activities are:

- Construction.
- Planning for startup (SP).
- Mechanical and electrical completion and precommission (SME).
- Commission, performance testing and post-commission management (SCPP).

Goals for availability engineering and management during construction. Availability engineering and management is concerned with the following during construction:

- Confirming the quality of plant equipment installation with respect to the points of sensitivity for reliability performance.
- Installation of availability equipment and elements such as handling equipment and shop machines.
- Acquisition of support materials and items such as spare/repair parts, testing equipment, etc.
- Timely identification, evaluation, and response to field changes.

Goals for availability engineering and management during the startup phase. Availability engineering and management should be concerned with the following during the startup phase:

- Confirming compliance with reliability, maintainability, and availability requirements and specifications.
- Planning, tracking and monitoring the deployment of maintenance operation functions, systems, and their related training operations.
- Capitalizing on opportunities to test and evaluate the availability scheme and its elements in the field under pressure.

Organization of Part Five

Part Five of the book has been divided into two chapters addressing the planning and execution stages of startup. Figure 15-1, as indicated in this chapter presents the first two stages: construction and startup planning.

Construction

It was mentioned that the construction process would not be treated extensively because of its association with availability engineering and management. This is not to imply that the discipline is not active during construction. However, its activities are most generally associated with startup.

The Goal of Construction

The fundamental goal of construction is to be ready for plant startup at an established date. This is partly accomplished by timely and effective procurement of materials. It is also accomplished with a successful transition between two modes of action. Initially, the mode of approach is most expedient for the general progress of construction. However, the approach must eventually shift to achieving scheduled startup milestones. In other words, construction is centered on these milestones rather than arriving at completion in mass on a single date.

Transition is necessary because construction will for a time progress according to opportunities to complete work. Schedule network approaches may be formulated, but the fundamental objective is to progress as opportunities arise during the construction process.

This approach becomes increasingly less appropriate, however, as construction progresses toward startup. It becomes important to transition into a focus on very specific startup milestones instead.

Consequently, the construction process will become increasingly disciplined. It will work toward bringing plant subsystems to completion. Timing will be governed by scheduled dates and sequences for testing subsystems in the startup stages.

Availability Engineering and Management During Construction

Availability engineering and management is active during procurement and construction. Its interests are as follows:

- To be made aware of any changes allowed in the manufacturing of plant equipment. These may come about as equipment is being manufactured. Availability management will be made aware of these changes through the procurement and associated quality assurance processes.
- To confirm points of installation that will affect an equipment item's reliability. These will have been identified by FMECA in the basic design phase. Such points include foundations, equipment anchors, structures, pipe bracing, etc. Precision alignment will also be a reliability issue. If installed improperly, these items create stresses on equipment that will greatly decrease its life and time-between-failures.

  Confirmation of reliability-centered installation issues is extremely important. Nonconformance can literally give the plant an availability performance for which the overall availability scheme has not been designed.

- To be notified if the siting of equipment, its surroundings, or its environment has changed. The project change and review processes will make availability management aware of such cases.

Planning for Startup

Planning for startup may begin along with the initial project management and control planning. There are two dimensions of availability management in the planning stage. They are:

- Planning for the many maintenance operation functions and systems that will be deployed as needed during the startup stages.
- Planning for the testing, evaluation, change, and improvement functions that support the startup stages.

Planning is crucial because safety management and control is more difficult than normal. Startup is dealing with a large number of subsystems that have yet to be tested as a system.

If this is not daunting enough, equipment failures are more likely at this point than any other time in the plant's life. Meanwhile, the operating and maintenance responses to failures have never been tested in terms of actual field conditions, the subject organization, and the personnel team. Thus, planning is critical even though the team may have experience in a similar startup.

At no other time is the plant faced with so much uncertainty. Thus, the best way to insure a safe and effective startup is planning. Furthermore, it should begin in the project design phases.

The following discussion identifies general planning requirements. It then defines the associated role of availability engineering and management.

The network of deliverables or activities for planning are coded as "TSP" (traditional deliverables in startup planning) in Figure 15-2. Each is discussed in the following sections. Figure 15-3 flowcharts the availability engineering and management deliverables. Its deliverables are coded as "ASP" ("availability" deliverables in "startup planning"). They are extracted from the description of traditional deliverables.

TSP1. Establishing Contractual Requirements

The contractual requirements of startup are established early in the project. They may be part of the overall engineering and construction contract.

Of course, there are many aspects to the startup contract. It is not the intent to explore them intensively here. The most important to a successful

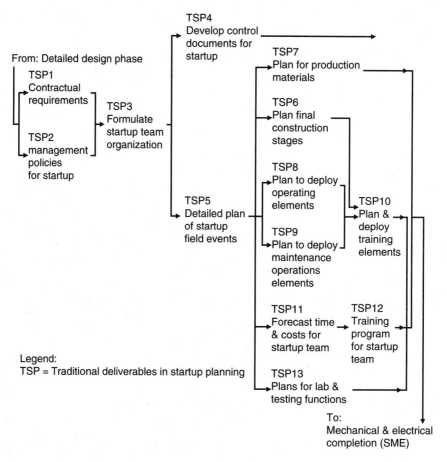

Fig. 15–2 Deliverables for the planning stage of startup (SP of Figure 15-1).

startup and, therefore, availability engineering and management, are the following:

- The responsibilities of the owner, contractors, and suppliers should be carefully defined. This is necessary because:

    Plant startup may be done by the owner with little outside help.

    Startup may be done with significant help by a project contractor.

    Project contractors for engineering and construction may be separate entities.

    Equipment manufacturers may also have responsibilities in accordance with the purchase contract.

    The engineering and construction contract may be turn-key, lump sum or cost plus.

- Contract terminology is a sensitive issue. Definitions for the same terms are often different. This is especially true for projects with international participants. Furthermore, the terminology will vary from one organization to the next.

  An example is the definition of mechanical and electrical completion. Construction completion may have a confusingly similar meaning. Another example is that mechanical and electrical completion may be defined as a milestone in some cases and as a stage of startup in others.

  Consequently, there is no right or wrong terminology. There is only agreed upon terminology.

- The transition events between startup stages must be carefully defined. This is especially true for the mechanical and electrical completion and precommissioning stages. The last stages of construction will still be underway just as they are becoming increasingly active.

  Various activities for certifying compliance with contract terms will be required before construction can be certified as complete. This will be tied to the definition of startup and milestones.

  Different entities will be responsible for these stages and milestones. Thus, exact terminology used in the conduct of their duties is a fundamental requirement.

- Construction and startup relationships must be defined. Different entities will need to integrate the conduct of their separate but related responsibilities. These responsibilities and relationships are identified with respect to phase completion. Completion will be formalized by certification and will involve payment requirements and transfer of ownership.
- The contract for startup should identify and define the documentation that will be necessary to all stages of the startup process.

ASP1. Availability management in the processes to establish contractual requirements (Figure 15-3). Availability engineering and management is concerned with the contractual aspects of plant startup. Like any discipline involved in the plant acquisition process, it must familiarize itself with the startup contract. However, the issue of contractual terms and details may go deeper.

The availability discipline is a new field to industrial plant design. Thus, availability management must be especially diligent in the determination that its needs are included in the contracts.

This is a significant challenge. Contract writers are likely to have a poor grasp of the availability field, its requirements, and their relevance to all contract parties. This is not the case for the traditional disciplines.

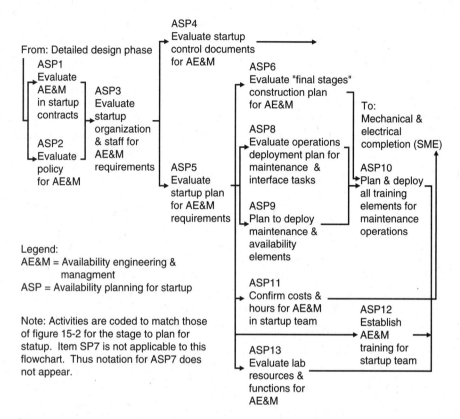

Fig. 15–3 Activities for availability engineering and management in the stage to plan for plant startup (SP of Figure 15-1).

TSP2. Management Policies and Decisions for Startup

Management must establish basic policies for project design and approach. Many are established during the plant feasibility study and conceptual design phases.

When planning for startup it is important to evaluate these policies. The most significant policies are:

- The fundamental responsibilities and roles of various entities in startup. These include availability engineering, maintenance, etc. It is otherwise possible that the contribution and benefits of important participants will be diminished. They may even be eliminated. This may be due to politics, short-term vision, or parochial perspectives.
- The degree and specifics to which the startup team will be involved in deliverables and their review during the plant design phases.

- The liaison between corporate, operating company, and plant management and the startup team. The objective is to avoid loss of coordination between their functions. It is also to manage the time pressures that may be placed on the startup team that can cause the principles of the entities in startup to be compromised.

  The policy is necessary because no plant or capital project is isolated: it is a venture within an owner's larger production and business system.

- Decisions for partial commissioning. These are important for their impact on existing operations, safety and hazards control. There are potential risks to safety and productivity from operating some parts of the plant while finishing others. Thus, partial commissioning creates additional layers of complexity in startup.

  Top management must take responsibility for this decision.

- The basic policy for quality assurance and control. The consequences of such policy are far reaching and relevant to the plant's entire life cycle. This includes plant design; procurement of equipment for designing specifications; construction, installation and testing of equipment; and plant functioning.
- The degree to which availability engineering and management will be included in the design, construction, startup, and commercial production phases.
- The policy for and basic approaches to process safety and environmental management. This policy is germane to the policy decision for availability engineering and management. Safety and environmental management has a substantial dependence on the availability discipline. This complex subject is beyond the scope of this book.

ASP2. Evaluate startup policy for availability requirements (Figure 15-3). Without the previously stated policies, crucial plant design, construction and startup activities may or may not occur. If they do occur, they may not receive appropriate and adequate treatment. Accordingly, availability engineering and management must evaluate startup policy for the following:

- Is there a basic recognition and policy for its inclusion and requirements? This is especially a concern if availability engineering has been brought into the project after the contracts, project plan, or plant design have begun to progress.
- How is the availability discipline specifically provided for in each policy issue? This should be carefully addressed since each policy must evidence management's decision to incorporate the availability discipline.

TSP3. Formulate and Implement Startup Team Structure, Responsibilities, and Authority

Startup is a difficult project challenge. The organization formed for its purpose will greatly decide the outcome.

Chapter 13 presented a design framework for organization design (AD8). The theory and design of any organizations, including that for startup, is a large subject. This deliverable will not again broach the subject.

The process for designing and planning the startup organization will incorporate the following in the principles and process of Chapter 13:

- Detail the startup process as a network of functions, deliverables, and their tasks. This is prescribed by Chapter 13, but bears repeating. The objective is to focus on the startup process. This task draws away from departmental and functional demarcations. These could be inappropriately carried over to the startup organization.
- Assess the human resources that are potentially available to the startup operation. A central consideration is the impact of diverting personnel from the functions that already depend on their talents. This activity will identify the need to evaluate alternative approaches in terms of cost, availability, and consequences.
- Establish policy in terms of owner and contractor sourcing, allowable time demand on personnel, concurrent duties, human resource management policy, etc.

  Human resource management policy is concerned with the careers of startup personnel. It is an important issue since some people will be taken outside their normal career path. They could be left behind or set back in their career and income.

- The assigned personnel should formulate the details of their position in the startup organization. They should then negotiate the detail with the other members.
- Establish plans for interface between production and construction shifts; plant personnel and the startup team; control room and field personnel; etc.

ASP3. Evaluate the startup organization for effectiveness in availability management (Figure 15-3). Availability engineering and management is deeply concerned with the design of the startup organization. Its effectiveness and results will suffer if the organizational concept and details are based on flawed logic. Therefore, the objective is to assure that the startup organization concept, structure, responsibilities, authorities, and relationships can effectively serve its needs.

An important concern is personnel assignment. Availability management must see that "its people" are on the team. It must assure that certain members recognize availability engineering in their constituency. These members must be brought to understand how their responsibilities, authorities, and actions serve availability requirements.

Therefore, the term "its people" means that there are representatives among the team of specialists of particular interest to availability management. They are as follows:

- The engineering design representatives. This may be one or several individuals responsible for all engineering disciplines. Availability and its various engineering fields is one such discipline that will look to the design representative to support its need to be notified and included in procedures for evaluating design and field changes.
- The maintenance representative. The maintenance operation functions are responsible for the results of the design phase deliverables for designing field and support aspects of availability performance.
- The safety management representative. Availability engineering as a discipline is crucial to the comprehensive treatment of production process safety management.
- The training representative. Maintenance and operator training is crucial to achieving the plant's availability goals. Recall that operator personnel typically have maintenance task assignments.

The evaluation and indoctrination of these representatives is integral to the evaluation of the planned startup organization. Availability management must evaluate whether the organizational scheme enables these representatives to effectively serve the discipline as a constituent.

TSP4. Develop Control Documents for Startup

Startup requires a great deal of documentation to guide, document, and certify activities and milestones. Much of it will ultimately become part of the plant's overall change and data management functions.

There are four fundamental requirements for controlling documentation: identification, preparation, application and management. Accordingly, the general types of control documents for startup are the following:

- Checklists for confirming that components are properly installed once plant support functions and elements are in place; and for guiding the various testing, evaluation and startup steps; etc.
- Documents to control, track, and manage the startup process.

- Documents to verify and approve or reject the results of activities.
- Procedures to be followed in the application and management of these types of control documents.

ASP4. Evaluate startup control documents for availability management require-ments (Figure 15-3).   The availability engineering discipline, like all other disciplines, must be concerned with the following:

- The adequacy and detail of the control documents with respect to availability issues.
- The design, construction, and startup functions that will administer them.
- How the control and change functions will make the availability discipline aware and a proactive participant in the need to review and revise the control documents. Availability management must assure itself that change management functions will be efficient, timely and expeditious.
- Whether the people associated with availability (see TSP3) on the startup team understand the principles behind the requirements of the control documents. One way to insure this is to educate those who will be associated with fulfilling them. Otherwise, some of these same members may not understand their importance. This suggests another role of the availability discipline during startup. It is to assure that the startup team is adequately educated and sensitive to the field of availability performance.

The control documents associated with availability engineering and management may begin to be developed during the basic design phase. At the very least, they will be developed as part of the plant design control documents for reliability and maintainability (AB1, Chapter 5). Ideally, the owner or project contractor will own baseline documents. If so, they must be adjusted to suit each startup.

TSP5. Develop a Detailed Plan and Schedule of Startup Events

A detailed plan and schedule of startup events should be formulated. It is derived from target dates associated with final performance testing and full commercial production. From the target dates, a network schedule is developed by working backward through commissioning, precommissioning, and mechanical and electrical completion of each subsystem.

Planning should also include activities to identify potential trouble spots that result from technical difficulties, unknowns, and elements for which there is limited experience. The objective is to develop contingency plans for the detailed plan and schedule.

ASP5. Review the master startup plan for availability management requirements (Figure 15-3). The planning and scheduling of startup events is important input to the planning process for availability engineering and management. The primary inputs and interests are as follows:

- The details required to plan the deployment of availability management and maintenance operations.
- Input to both manage and confirm the appropriate attention to control documents for availability requirements. The master plan provides a basis for formulating a checklist of necessary control documents.
- A basis for monitoring, testing, evaluating, and deploying availability and maintenance operations in juxtaposition with the installation and startup activities that require them.

TSP6. Plan for the Final Stages of Construction

As mentioned, construction will progress considerably towards completion with approaches that are maximally beneficial to the construction process. During that time, activities may appear disjointed and random. However, as a complex and somewhat chaotic undertaking, the ruling objective is to progress without loss of quality.

The ruling objective will change at some point in the progress of construction. It will be to channel the remaining construction events into a rigorous schedule of events for mechanical and electronic completion, precommissioning, commissioning, and startup.

The previous deliverable developed the detailed schedule of startup events. This deliverable is for extending that plan to a detailed schedule of activities in the final construction stages. The time for the transition between ruling objectives will become apparent in the planning process as it spans both deliverables.

ASP6. Review the construction plan for availability management requirements (Figure 15-3). Availability engineering and management is especially concerned with the construction plan and schedule. This is especially so as the second objective begins to rule.

Availability management may wish to identify, plan, and organize to use equipment now in place to validate important maintenance procedures. This may be the case during both the chaotic and the channeled stages of construction.

Availability modeling during the design phase identified where plant availability and safety is most sensitive to the quality of specific maintenance tasks. These findings may lead to special testing and evaluation during construction.

TSP7. Formulate Plans for Manufacturing Materials and their Management

Planning for startup should include storage and handling of manufacturing materials and the produced product. This requirement is not significantly germane to availability engineering. It is included here for the purpose of presenting a complete picture of the startup process.

TSP8. Formulate a Detailed Plan to Deploy All Operations and Support Elements

Plant operation and functions, systems, and resources are deployed in a planned and organized manner. Timing must mesh with the final construction stage and the subsequent startup stages.

The fundamental objective is to assure that permanent personnel are involved in startup and preparation for their permanent duties.

ASP8. Evaluate operations deployment for assigned availability, maintenance, and interface tasks (Figure 15-3). Literature and plant designers are often too quick to include operating and maintenance activities in the term operations. Both are often defined according to the department in which the task is assigned rather than its description.

This is not a useful approach. Operations and maintenance tasks are often assigned across plant departments. Thus, availability management will confirm that such crossover assignments are appropriately treated in the overall deployment plan for operations.

Returning a failed piece of equipment to service involves the interface of operating, maintenance, and manufacturer activities. Availability management must evaluate the deployment plans for the support functions of such interface tasks.

This and the equivalent deliverable for maintenance operations should be developed as an integrated deployment plan. To do well in one regard but poorly in another reduces the success of the overall startup.

TSP9. Formulate a Detailed Plan for Deploying the Availability and Maintenance Operations

Availability and maintenance operation functioning was detailed during the plant design phase. The detail included

- Maintenance tasks and procedures.
- Determination of human and physical resources including staff, skills, parts and components tools, etc.
- Maintenance operation support and management functions.
- A training program.
- Availability management functions and systems.

The well-planned deployment of availability management and maintenance operations is a key to the plant's initial financial performance. Along with the operating functions, it affects the time and cost required to achieve startup and subsequent full commercial production. Thus, the deployment plan and schedule must establish the appropriate timing for putting availability functions, systems, and resources in place.

Deficient functional deployment and resource levels will undermine the startup. There will be a poor response to the natural chaos and failures that are typical to the first hours of operation.

Excessive deployment has drawbacks beyond financial waste. It may distract the team from its purpose. This is because excess people, material, and equipment have to be managed.

The requirement to plan for deployment is shown as TSP9 in Figure 15-2 and as ASP9 in Figure 15-3.

TSP10. Plan and Deploy All Training Elements

The timely deployment of the training program is integral to the deployment of operating and maintenance elements. It is necessary to begin the training program with respect to skills needed for startup and commercial production. The needs and programs for operating and maintenance are carefully formulated during the detailed design phase.

Training can be distinguished for the needs of the following two overlapping groups:

- The startup team of temporary and permanent plant staff.
- The personnel to become permanent plant staff.

Training for the startup team is treated separately (TSP12 and ASP12). This is because the basic tasks of these two groups are different.

However, there are common requirements for both groups. They include general induction, safety, plant layout, detailed operations, and emergency procedures.

The requirement for planning and deploying the training program is shown as TSP10 in Figure 15-2 and as ASP10 in Figure 15-3.

TSP11. Forecast the Cost of and Time for Startup Human Resources

Startup is a substantial organizational activity and cost. Its stages can span several years. Thus, a resource, cost, and time estimate is prepared for the startup team.

The various cost estimates may have been drafted during the conceptual design phase. If not, it may be part of the engineering and construction contractor's project execution plans. This deliverable will test the assumption of such earlier costs and schedule.

From early and evolving project schedules, startup management must determine when the startup team must be on-site and for how long. However, the analysis of the costs and schedule is not limited to the startup team members. It must include the human resource requirements for periodically supporting the team and its functions. It must also assure that the construction estimates include the necessary crafts.

ASP11. Evaluate startup human resource cost and time for the needs of availability management (Figure 15-3). Availability management must be concerned that the resources, cost, and time estimate for the startup team include its needs. This is critical. Once a ship has sailed, it is very difficult to get aboard. Or if aboard, to change its bearings.

Protection against such a possibility begins during the earliest project planning phases. The initial project planning may have included tasks and budgets for all disciplines through startup. If this is the case, the availability discipline will have confirmed that its needs were understood and translated to tasks.

Chapter 4 introduced owner activities to assure that availability engineering and management requirements are identified, tasked, and budgeted. The owner has a responsibility to make availability engineering requirements initially known and understood. It must later confirm that internal and external entities involved in the project phases have responded to these requirements. Otherwise, it will be difficult to later gain the attention and sincerity of overall project personnel.

There are many consequences from failing to review the startup team plan. None of them is good. One is that the activities of the design, maintenance, safety management, and training representatives may not be well integrated with the availability discipline. Instead, the now limited human resources and budget will disallow the time and energy required to integrate the availability management processes.

TSP12. Determine, Develop, and Execute Training for the Startup Management Team

Startup is a complex, dynamic management problem. Discussion in Chapter 13 showed that rising to a complex challenge has fundamental organizational design requirements, which are the significant cross-organization functioning and information systems. This principle is germane to the startup organization.

Chapter 13 showed that a divisional structure for startup means that the startup problem must be made simpler. This may not be an acceptable alternative unless management is willing to use slack resources (i.e., greater personnel, more funds, longer time-to-complete) to achieve the desired result.

A key to lateral strategy is to train the startup team members in all disciplines pertinent to startup. Thus, they will become specialists guided by a strong sense of what their colleague disciplines are about. In other words, training will develop lateral knowledge to complement the specialized knowledge of each team member.

ASP12. Determine availability training for the startup team (Figure 15-3). Availability management has special concerns for training the startup team. One is to assure that the people associated with availability performance are made knowledgeable.

This is especially important. Those touched, involved, or served by the discipline may have little sense of such cases. For example, the maintenance representative may see her or his role as only providing maintenance rather than delivering availability performance.

The most likely categories of training are as follows:

- Some design and management discipline representatives are instrumental in the objectives of the availability discipline. Consequently, they are in some degree practitioners. They may require rigorous training of up to five working days.
- General team membership will need a reasonable grasp of availability management requirements and their purposes. This may be accomplished with one or two days of training.
- The same training may include individuals on the outward side of the liaison positions to the startup operation.

TSP13. Formulate Detailed Plans to Deploy Performance Testing and Laboratory Functions

The schedule and quantity of testing and laboratory functions and resources are derived from the following sources:

- The detailed startup event schedules.
- The deployment plans for operating, maintenance and training functions.

ASP13. Confirm adequacy of laboratory resources for availability performance support needs (Figure 15-3).　The availability discipline must confirm that its testing and laboratory requirements are planned for deployment.

Bibliography

Construction Industry Institute. *Planning Construction Activity to Support the Start-up Process.* Austin, TX. 1990.

Harrison, Roger. Startup: *The Care and Feeding of Infant Systems.* Organizational Dynamics. Summer 1981.

Horsley, D.M.C. and Parkinson, J.S. *Process Plant Commissioning.* Institution of Chemical Engineers. Rugby, Warwickshire, England. 1990.

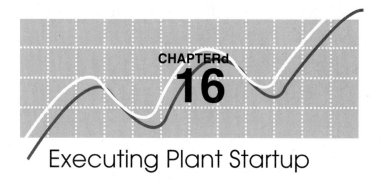

CHAPTER

16

Executing Plant Startup

Introduction

The previous chapter explored the general process of planning for startup. This chapter introduces the sequences of executing these plans. The bounded activities shown in Figure 16-1 are the subject of this chapter. They are:

- Mechanical and electrical completion and precommission (SME).
- Commission, performance testing, and postcommission management (SCPP).

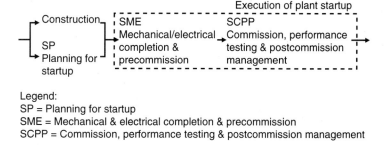

Legend:
SP = Planning for startup
SME = Mechanical & electrical completion & precommission
SCPP = Commission, performance testing & postcommission management

Fig. 16–1 Stages in the execution of plant startup.

The approach of the previous chapter continues and a network or flow-chart of startup activities is introduced. The needs and objectives of availability engineering and management associated with each are then identified. This identification is used to form a flowchart of availability engineering and management within the bounded activities of Figure 16-1.

Mechanical and Electrical Completion and Precommissioning

The importance of defining contractual terms applicable to startup was discussed in Chapter 15. The definition of mechanical and electrical completion and precommissioning is such a case.

Literature varies in the definitions of completion and precommissioning. Mechanical and electrical completion is sometimes defined as a milestone. Other times reverse definitions are given to completion and precommissioning.

Therefore, it is still necessary to define mechanical and electrical completion and precommissioning. For this book they are as follows:

- Mechanical and electrical completion attempts to assure the following:

  Equipment is installed correctly.

  Equipment works correctly.

  Instrument and control systems work.

  Plant components and subsystems are ready for precommissioning.

- Precommissioning is the preparation of plant subsystems for commissioning as a system. Its primary objective is to discover and eliminate problems that would otherwise appear at a more critical stage. Accordingly, it involves

  Tests such as simulations, hydrotesting, and equipment rotation.

  Discerning that all plant functions and procedures are in place.

  Assuring that production materials are in place and ready.

The deliverables for mechanical and electrical completion and precommissioning are coded as "TSME" (traditional startup activities for mechanical and electrical completion and precommissioning) in Figure 16-2. Figure 16-3 flowcharts the associated availability engineering and management deliverables. Its deliverables are coded as "ASME" (Availability startup activities for mechanical and electrical completion and precommissioning).

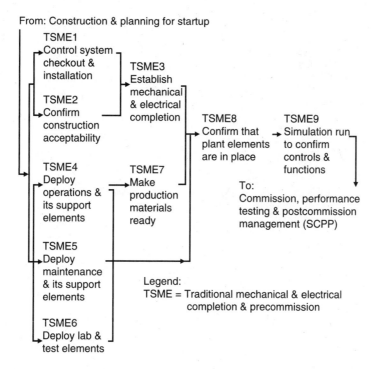

Fig. 16–2 Deliverables for mechanical and electrical completion and precommissioning stages of startup (SME of Figure 16-1).

TSME1. Checkout, Installation, and Commissioning Instrumentation and Control System

Mechanical and electrical completion and precommissioning of the instrumentation and control system is a special problem. Testing of the instrumentation system and the control system is often done in separate locations. The instrumentation system cannot be installed and checked until the plant is close to being completed. Meanwhile the control system is tested at the supplier's facilities.

The controls still must be precommissioned as a system. Therefore, the instrumentation and control system is actually commissioned as the remaining plant is still being precommissioned.

The requirements to check, install, precommission and commission the integrated instrumentation and control systems are as follows:

- Test and calibrate control system hardware and software at the supplier site.
- Install the control system at the plant site.

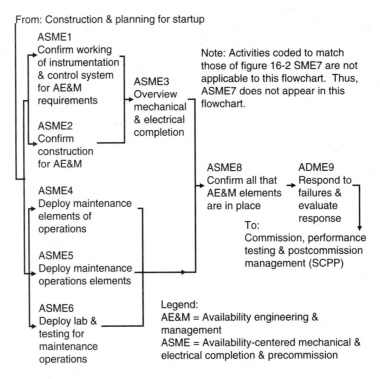

Fig. 16–3 Activities for availability engineering and management in mechanical and electrical completion and precommissioning stages of startup (SME of Figure 16-1).

- Confirm that the infrastructure for instrumentation is installed and able to support testing. Infrastructure includes power, air, cables, tubing, etc.
- Check the instrumentation system for installation, connections, leaks, etc.
- Calibrate instruments for range and bias.
- Function test the instrumentation and controls as a system.

ASME1. Confirm that the instrumentation and control systems will serve the role of providing data to availability management (Figure 16-3). Availability management has concerns for the instrumentation and control system. It is a source of reliability data. It may also have a role in the scheme to monitor and detect imminent failures.

The testing and evaluation plan for availability engineering will have identified specific aspects it wishes to confirm during these startup activities. These will be related to the reliability and monitoring data to be collected. It will also wish to confirm that the systems and elements that transfer data to the availability-centered improvement, change, and data management systems can perform as planned.

TSME2. Confirm Acceptability of Installation and Construction

The objectives for confirming the acceptability of equipment installation and plant construction are as follows:

- To determine if there are variations between the baseline design (construction documents) and the actually constructed and installed condition (as-built documents).
- To evaluate these variations or confirm that they have been evaluated.
- To verify that the as-built conditions have been incorporated into the plant baseline design documents and data and that they have found their way into the improvement, change, and data management functions and systems.

ASME2. Availability-centered confirmation of acceptable installation and construction (Figure 16-3). Availability management is concerned with confirmation for the following reasons:

- To determine if there are as-built conditions pertinent to availability performance.
- To evaluate changes against the current baseline availability design.
- To decide whether these changes will require that the testing and evaluation plans for availability performance be revised.

The second case attempts to determine whether plant reliability, maintainability, and availability characteristics and economics have been affected. The review will discover if a design change is needed. Such changes are either to reverse the as-built condition or counter it. This can include the realignment of maintenance operations.

Therefore, availability management is concerned with the earliest possible identification of any divergence from baseline design. Options to change the plant's design will become increasingly limited as time passes.

TSME3. Determine Mechanical and Electrical Completion of All Subsystems

Mechanical and electrical completion of all plant subsystems is established. They include production, ancillary, utilities and materials handling subsystems. Previously developed checklists and other control documents and procedures are applied to make this determination.

ASME3. Availability-centered overview of established mechanical and electrical completion (Figure 16-3). An availability-centered review of completion will have the following focus:

- To participate in the testing and evaluation of requirements if there is an availability-centered interest. Such interest will have been determined during the plant design phases.
- To determine if in the act of achieving mechanical and electrical completion, there were changes in equipment or the subsystem of which they are part.

TSME4. Deploy Operating and Support Functions and Systems Elements

The deployment of operating and associated support functions and systems was planned in a previous activity (TSP8, Chapter 15). At this time, their planned deployment should be drawing to a finish.

ASME4. Availability management: Participate and monitor deployment of plant operating elements (Figure 16-3).   Availability management is concerned with the deployment of operations and its support functions. The availability scheme will have planned that some maintenance tasks and roles will be assigned to operating functions. Thus, availability management must confirm that these roles are treated in deployment.

Another concern is the points of interface between maintenance and operating functions. One is the equipment tagging and tracking system which is the interface where the actions of production process operations influence time-to-maintain. These interfaces can also be the cause of an accident during maintenance.

Therefore, availability management may be involved directly in the deployment of such crossover elements in the availability scheme.

TSME5. Deploy Maintenance and Support Functions and Systems

Planning to deploy the maintenance operation was the result of the stage for planning for startup (ASP9, Chapter 15). The deployment of the functions and elements of availability performance should also be drawing to a finish. This is especially so for the maintenance operation.

As part of deployment, there may be support requirements during mechanical and electrical completion and precommissioning. The subject equipment may require maintenance functions to respond to a failure.

These are important opportunities. This is because more than just plant equipment should be subjected to test and evaluation during startup. The maintenance operation functions and elements are also subject to scrutiny as these opportunities arise.

This requirement appears identically in Figures 15-2 and 16-3 as TSME5 and ASME5.

TSME6. Deploy Testing and Laboratory Functions and Elements

Laboratory and testing requirements are an integral part of plant start-up and subsequent commercial functioning. They are applicable to the following:

- Testing incoming materials.
- Quality testing of the produced product.
- Equipment condition monitoring and acceptance for assuring compliance to specified mechanical performance.

At this point, the deployment planned earlier (TSP13, Chapter 15) is drawing to a finish.

ASME6. Availability management: Confirm deployment of pertinent testing and laboratory provisions for availability performance (Figure 16-3). Availability engineering must be concerned with elements that are part of the availability scheme. These are associated with requirements such as testing lubricating oils and other plant materials and conditions related to reliability.

TSME7. Make Production Materials Ready

The operation functions have been deployed or are now late in the process of being deployed. One milestone is making the plant production materials ready for simulation and final commissioning. This activity has minimal relevance to the availability discipline. It is presented here for completeness.

TSME8. Final Confirmation that All Systems, Materials and Operating, Maintenance, and Administration Functions and Elements are in Place

Before commissioning, it is necessary to make a final overall assessment to determine that all necessary materials and plant operating, maintenance, and administrative functions and elements are in place. This includes the availability and maintenance management information system and its ability to begin collecting equipment and maintenance task detail history. The functions and elements of quality assurance and control, safety management, etc., must also be confirmed. Assessment should also verify that the required training has been accomplished.

ASME8. Confirm that all necessary availability functions and elements are in place (Figure 16-3). Availability management should both involve and subject itself to the confirmation process and requirements presented in the previous section.

TSME9. Simulation Runs for Confirming Function and Controls

Simulation runs involve plant components and subsystems. The objective is to confirm that they are functioning correctly. Depending on the contractual terms for startup, this may be the first time the operations and maintenance functions and systems are called into play. In other cases these tests may have been part of mechanical and electrical completion activities.

ASME9. Availability management: Respond to failures and evaluate maintenance response during subsystem simulation (Figure 16-3). Equipment failures may occur during these tests. When this happens, the operators should isolate equipment for maintenance. Maintenance will be called upon to return the failed item to availability.

At that point the operators will remove the item from isolation. This will present opportunities to evaluate the maintainability dimension of the availability scheme under stress. What needs to be determined is whether the proactive plans for maintenance response to failures can be carried out as expected.

Commissioning, Performance Testing, and Post-Commissioning

The definition of the term commissioning varies. Some define it as testing plant subsystems. When this is completed, the plant is ready for the introduction of raw materials and any associated gases or liquids to the production process. Others define commissioning as making the plant live for normal functioning. The operating parameters are then adjusted to attain the specified process performance. The second definition will be the one generally applied in the following discussions. The first definition comes closer to defining the process of commissioning described above.

The combined deliverables for plant commissioning, performance testing, and post-commissioning coded as "TSCPP" (traditional startup activities in commissioning, performance testing, and post-commissioning) are shown as a network of activities in Figure 16-4. They are each presented and

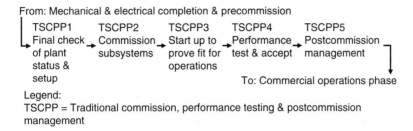

From: Mechanical & electrical completion & precommission

| TSCPP1 | TSCPP2 | TSCPP3 | TSCPP4 | TSCPP5 |

TSCPP1 Final check of plant status & setup → TSCPP2 Commission subsystems → TSCPP3 Start up to prove fit for operations → TSCPP4 Performance test & accept → TSCPP5 Postcommission management

To: Commercial operations phase

Legend:
TSCPP = Traditional commission, performance testing & postcommission management

Fig. 16–4 Deliverables for plant commissioning, performance testing and post-commissioning stages of startup (SCPP of Figure 16-1).

described in the following section. A flowchart that represents the associated availability engineering and management process for startup is presented in Figure 16-5. Its deliverables are coded as "ASCPP."

TSCPP1. Final Check of Plant Status and Setup

The subsystems and associated equipment have been confirmed by inspection, simulation, and testing to be fit for their purposes. All plant functions and their elements have been trained and deployed. All equipment has either a status of readiness or is isolated from the system. The objective is now to make sure that the status of all plant elements is consistent with what has been reported.

ASCPP1. Declare status of availability management elements in the final check of plant status and setup (Figure 16-5). The availability management and maintenance functions and system must verify or declare the status of these elements. This is to confirm their capacity to respond as needs arise in the plant. The required response may be analytical processes or field actions. The status of the availability-centered change, data, and management functions and systems are core elements of both possibilities.

TSCPP2. Commission Subsystems

Initially, the plant subsystems may be commissioned individually rather than commissioned for the entire plant. This may be in the order of utilities, laboratory, input materials handling, ancillary equipment, stages of production and reaction, product storage, and materials handling.

The objective is to gain confidence in the subsystems as an operating system and in the operator's ability to control them. The strategy also recog-

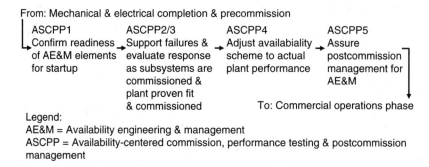

Legend:
AE&M = Availability engineering & management
ASCPP = Availability-centered commission, performance testing & postcommission
management

Fig. 16–5 Activities for availability deliverables in the commissioning, performance testing and post-commissioning of startup (SCPP of Figure 16-1).

nizes a much higher than normal probability of equipment failures during first hours of use.[1]

Thus, the divide and conquer approach confronts such needs and challenges in a less stressful situation. It also helps control the hazards inherent to commissioning a new production system.

ASCPP2. Availability management: Support failures and evaluate maintenance response as subsystems are commissioned (Figure 16-5). The maintenance operation will continue to gain experiences in the field. This is because many more than usual mechanical problems can be expected to arise. Many occur from infancy failures. They also occur when attempts to operate subsystems for the first time place unusual stresses on individual equipment.

Evaluation processes must be formulated as part of the living availability design scheme. These should be positioned to evaluate maintenance actions, procedures, and management functions as they are applied.

TSCPP3. Start up Plant to Prove Fitness for Operations

The plant is run to confirm that it is fit to fulfill its purpose in terms of technical performance and productivity. The nature and determinates of fitness are a function of whether the plant involves a batch or a continuous production process. Some plants may involve both.

The batch process. Fitness for a batch process plant is concerned with achieving the following:

---

[1] Some feel such a preconception must be eliminated. They look to availability design and implementation to progressively reduce such failures to a "theoretical" zero.

- The correct mixture of materials.
- Production sequences.
- Critical production parameters.
- Confirmation of the process and procedures for routine production.

The continuous process. In contrast, the fitness of continuous discrete or process manufacturing is concerned with achieving the following:

- Stable operating conditions.
- Confidence in the ability to achieve a reasonable running time without an upset.
- Confidence in the instrumentation and control system.
- Product specification.

ASCPP3. Capitalize on the opportunity to test the fitness of the availability scheme (Figure 16-5). This stage will continue to generate maintenance needs that will challenge and, thus, test the availability management and maintenance operation scheme.

The short-term capacity to test the plant hard design for availability performance is limited. Reliability and maintainability can only be tested as data and information are collected and analyzed over the long-term. This is the purpose of the living analysis capability of the availability scheme.

Therefore, determining fitness with respect to availability performance is only possible in terms of whether maintenance operations, field actions, and resources are appropriately provisioned, effective, and efficient as planned. Even this presents a limited number of short-term possibilities.

This limitation is a reason why the capacity to design and model plant availability is so crucial. Otherwise, there is no legitimate means to confidently establish an expectation of performance. Performance will only become somewhat apparent much later in the plant's life.

TSCPP4. Run to Performance Test and Acceptance

The objective is now to run and test the plant for design conditions, rate, and productivity. The requirement is to bring the plant to controlled steady-state conditions. It is then to achieve the specified process performance for a specified period of time. The following outcomes are possible:

- Achieving the goals.
- Overcoming the problems of achieving those goals.
- Falling short of goals with no possible means of remediation.

Adjustments will be required if the plant cannot achieve its performance goals. Individual or multiple changes may need to be made to the production process, plant functioning and resource levels. The plant design should be changed only as a last resort.

ASCPP4. Adjust availability scheme to fit performance test results (Figure 16-5). The maintenance operation scheme will offer possibilities for adjusting plant performance to meet its goals. Chapter 1 showed that reliability and maintainability are partially a function of the maintenance operation. Thus, plant productive capacity and, therefore, cost structure may be adjusted. Specifically, operational availability (A_o) in Figure 9-4 may be shifted with respect to achievable availability (A_a). This is accomplished by changing the maintenance intervals by changing human (including overtime hours) and material resource levels. Organizational effectiveness can also be revised. This is done by appropriate staffing and varying the power of organizational structures and processes.

One purpose of the improvement, change, and data management functions for availability management is to allow such determinations. They supports the reformulation of the availability scheme for most profitably matching the actual plant production process performance.

TSCPP5. Post-Commissioning Management

The startup process will generate data, changes, etc., that must be managed. Failing to manage these as a resource is a potent source of risk to all aspects of plant productivity and safety. Such risks are the worst kind because they creep unnoticed into the plant's being.

The general requirements of post-commissioning management are as follows:

- Final evaluation of modifications and organization of any follow-up activities.
- Filing of tests and other data in the appropriate data and change management systems. These are, in essence, baseline performance and initial settings.
- Plans and needs for post startup audits of performance.
- List of problems for follow-up.
- Confirmation that all data collection systems are in place and functioning as planned.
- Achievement of the initially planned levels of spare/repair parts, components, and material for normal commercial operations.

ASCPP5. Monitor availability-centered elements of post-commission management (Figure 16-5).   Availability management is deeply concerned with these aspects of post-commission management for the following reasons:

- Any plans for modifications and changes in response to the reservations list may change the system and, therefore, elements throughout the availability design.
- The startup test and setpoint data may be the initial point plant data that the living availability engineering process will draw upon.
- Availability management will influence the nature of planned audits. One goal of availability engineering is to test, evaluate, and improve the plant and, therefore, its own performance.
- Availability engineering and management is directly responsible for parts, components, and materials inventories. As such, it will be sensitive to the need to treat parts as a moving target as plant performance rapidly becomes more reliable.

Summary

This is the final chapter for discussion of the plant construction and startup phase. The general startup process was described. This was the basis to in turn describe one for availability management.

It was established that the full availability scheme should be in place at the end of startup. This includes the maintenance operation and the many overall availability management functions and systems.

The early stages of commercial life will challenge them because plant functioning will progress along a learning curve. Meanwhile, equipment failures will be greater as they decline to a normal level.

Therefore, availability management during startup is a crucial dimension of early success because it will help shorten the time to achieve the first production and target production levels. The success at this stage, however, is based on the success of availability engineering and management during the design phases.

Bibliography

Construction Industry Institute. *Planning Construction Activity to Support the Startup Process.* Austin, Tx. 1990.

Harrison, Roger. Startup: *The Care and Feeding of Infant Systems.* Organizational Dynamics. Summer 1981.

Horsley, D.M.C. and Parkinson, J.S. *Process Plant Commissioning.* Institution of Chemical Engineers. Rugby, Warwickshire, England. 1990.

PART SIX

Commercial Operations Phase

CHAPTER 17

Classic Cycles of Production System Management

Introduction

The availability discipline is established in the plant as a living process. Availability management will become part of total plant functioning. Otherwise, the ability of the plant to perform according to the most profitable availability scheme will generally decline over time. This chapter explores the role of the availability discipline in corporate, operating company and plant management. Subsequent chapters explore the role in detail as a set of work and decision processes.

Empowerment by Availability Engineering and Management

This exploration will reveal that it is a mistake to view availability management and its maintenance operations only in the context of plant functioning and improvement. Instead, availability management and its living processes are a substantial, natural and integral part of all classic management cycles. These cycles occur throughout the corporate, operating company, and plant organizations. All three are empowered by availability management to more closely achieve their full potential in creation of income and maximum use of assets.

This empowerment is made possible by the availability engineering work products of the previous design phases. There are now findings, tools, work processes and functions in place to serve the plant throughout its producing life.

Approach to Exploring Availability Management in the Plant's Producing Life

Writing should be an exploration. Discoveries occur as a writer carefully develops the logic that connects the ends to the middle. That was very much the objective and the case for writing this part of the book.

This exploration begins by finding a basis to describe the nature, needs, and business processes of the plant's productive life. This description must also regard the plant as part of a larger production and business system.

Thus, the first challenge was to step beyond merely describing the discipline in the plant's productive cycles. The answer would have been limited to maintenance operations and evaluating, improving, and adjusting availability performance.

Therefore, the points that are explored are the following:

- What are the universal classic production systems operations management cycles?
- How do they each trigger all types of change in the plant's design and functioning?
- What role does availability management and its living design capability play in those triggered changes?

Organization of Part Six

This chapter identifies and introduces those cycles. It then explores the role of availability engineering and management in these cycles. The next three chapters translate those roles to availability management processes within the classic cycles.

Classic Cycles of Production System Operation Management

Classic cycles of management can be distinguished when the plant is viewed within the owner's larger business enterprise. This begin with strategic cycles and continue through day-to-day on-line production cycles and subsequent performance evaluation.   The entire chain includes cycles of design, construction, and startup.   All occur both laterally and vertically along the corporate, operating company, and plant organizations.

The Classic Cycles

More specifically, the following management cycles will be introduced:

- Strategic and long-term financial planning beginning with the analysis of business and technical environments.   Strategies are formulated to capitalize on opportunities and to counteract threats.
- Application of research and development with respect to all aspects of productivity.   This spans product, production, and availability performance.
- Activities for developing and refining the plant.   Plant engineering, construction, and startup are integral to this cycle.
- Aggregate planning cycle.   The challenge of this cycle is to match the operating company's productive capacity and resources to forecasted demand.   It might be said that the concern is a business-centered planning for production.   Forecasts are for several years in advance.
- Operating budget cycle.   The purpose of this cycle is to explore and maximize the financial ramifications of the aggregate plan.
- Detailed production planning cycle.   The purpose of the planning function is to convert the near-term periods of the aggregate plan to production and resource management detail.
- Production cycles.   These cycles are a normal sequence of preparation for startup, startup, achieving steady-state production, shutdown, and turnaround to another cycle.
- Maintenance operations cycles that serve the production cycle.
- Performance evaluation cycles.   These are concerned with product quality, relationships of input to output, availability performance, budget, etc.

The Nature of Cycles

These management or business process cycles that occur repeatedly throughout the plant's producing lifetime are not a matter of beginning with the strategic cycle (TO1) and passing through each to the end of the overall production and business system operations management cycle and then starting over.   Instead, the plant's entire productive life will be cycles of change in functioning or design.   Each will reflect decisions by corporate, operating company, and plant management.

Clustering of Cycles

Figure 17-1 shows that the production operations management cycles of the commercial production phase fall within the following clusters.

- Long-term vision.
- Middle-term planning.
- Short-term planning and action cycles.

The production and pertinent business systems operations management cycles are coded as "TO" in Figure 17-1. This is to denote traditional production operation management cycles. They are described in the following sections.

Long-term vision. The first cluster of cycles has a long-term vision. It sets the direction and capability of the organization. Its cycles are

- Strategic and long-term financial planning.
- Research and development.
- Product, process, and plant development.

These cycles are concerned with the directions that can be taken, the knowledge and technologies that will be used to perpetuate or fulfill those decisions, and the development of the system to incorporate those technologies. Each cycle of the cluster can be triggered by the feedback from the other two clusters.

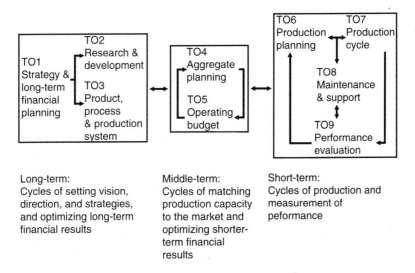

Long-term:
Cycles of setting vision, direction, and strategies, and optimizing long-term financial results

Middle-term:
Cycles of matching production capacity to the market and optimizing shorter-term financial results

Short-term:
Cycles of production and measurement of peformance

Fig. 17–1 Cycles of the production system operations management cycle.

Middle-term planning. The second cluster has a middle-term orientation, which pieces together the organization's resources to perform short- and middle-term production management activities. The aggregate planning and operating budget cycles are concerned with matching the organization's production capacity to short- and middle-term demand in the most profitable manner. The cluster will also provide influential feedback to the previous cluster.

Short-term planning and action cycles. A third cluster is concerned with using and fine-tuning production and organizational plans and elements. The production planning, production, and maintenance operation cycles are at the heart of that purpose. Ultimately, the performance evaluation cycle will assess how well the core cycles of production actually fit with what the strategic and middle-term cluster had intended. Therefore, the cluster provides feedback to the other two.

Disequilibrium as a Trigger of Cycles

Each cycle within a cluster is related to other cycles in its cluster. It is also related to those in other clusters. When one cycle arrives at a disequilibrium, others are subject to activation. Which cycles respond is a function of the nature of disequilibrium. Disequilibrium is defined as follows:

- The inability to fulfill a strategy.
- The inability to achieve assigned performance measures.
- A need for change in response to strategic challenges, or threats.

Therefore, a cycle can trigger others. Initially, the response may be an adjustment. Eventually, a small to large cycle of design, construction, and associated startup will occur.

The plant may be just one in a larger production system. These same cycles in related upstream, downstream, or parallel production facilities can trigger a cycle in a subject plant. The reverse is also true.

TO1. The Strategic Cycle

Corporate, operating company, and plant management must develop long-term strategies. This is typically an annual cycle in modern-day business management.

Hierarchial Levels of Strategy

There are hierarchies of strategy parallel to the nature of corporate, operating company and plant management. Each has a different focus. Combined, they form a single holistic strategy for the total organization. The scope of each hierarchial level is as follows:

- Corporate strategy is focused on which businesses to operate, allocating resources to those businesses, and achieving competitive advantage through their synergy. Its functional focus will be on financial performance and organizational structure.
- The primary strategic issues of the operating company are achieving distinct competencies and consequent competitive advantages for particular industries and market segments. They will reflect the stage in product and the market life cycles. Synergy is focused on the integration of business and production functions and processes.
- Plant strategy is focused on achieving maximum plant productivity with respect to required and possible productive capacity.

Possible Strategic Decisions

The strategic cycle has two stages. The first is analyses; the second is strategic decisions. The decisions include many possibilities. Some are as follows:

- Product line and associated quality with respect to the customer. This includes the addition of new products and changes to existing products.
- Policies for quality and productivity.
- Matching the productive capacity of the owner's system of plants to market demand in a way that produces the best business performance.
- The allocation and level of human and financial resources.
- Basic policies for responding to forecasted long-term demand with existing productive capacity.
- The decision to capitalize on research and development with respect to products, production processes and equipment, materials, management methods, customer needs, and market behavior.

Availability engineering and management is critical to all of these decisions. The discipline has the following relationships:

- Availability management is affected and empowered by the decisions and their relationship to availability performance.
- Management is empowered to much better explore, define, and make these classic decisions.

Long-Range Financial Planning

The strategic cycle includes long-range financial planning. This is the analysis of the consequences of candidate and chosen strategies on financial statements for many years to come.

Thus, the strategic cycle will consider the owner's long-term financial picture. This view looks at the case for "what if" the company continues as it is and assesses the financial consequence of the strategies being evaluated and considered.

Management is often regarded as having a short-term perspective. This is sometimes true. It is also because employees and the public do not often see the long-term financial analysis process in action. Instead, they observe and are more widely involved in the short-term operating budget process. Its purpose is to maximize the short-term with respect to long-term strategies and financial expectations.

Detailed Strategic Plans

Ultimately, the cycle selects strategies and prepares detailed plans for their realization. The strategy will establish the following:

- The scope of the organization's business.
- Resources to be deployed across that scope.
- Competitive advantages to be achieved.
- Synergy that is expected to create or advance that advantage.
- Organizational restructuring consistent with the previous elements.

Role of the Availability Discipline

Availability engineering and management has an important role in the strategic cycle. This is true across corporate, operating company and plant management. The primary nature of the contribution is as follows:

- To gather and assess availability data and information. This is integral to the strategic analysis of the organization's assets. Assessment is made in terms of their current nature and potential. This is part of the larger process to describe the organization's current and possible strategic position.
- To determine the capacity and means for changing the current availability scheme and performance. The issue addresses how to best serve the various strategies being formulated, considered, and ultimately selected.
- Availability performance is fundamental to financial performance. Thus, its capital and functioning costs are also a subject of long-range

financial planning. The living availability design should be used to study the financial ramifications for each strategy. This study will be concerned with long-term income and productivity of working and capital assets.

- To participate in developing the details of defining the final strategic plan. The top-level detail of availability performance is a major part of a strategy's description.

It is apparent that the strategic cycle is less effective without availability management as an organizational capability. Otherwise, a major dimension of the production system is not competently included and treated in the strategic process.

This has immense ramifications. Studies show that companies that plan do much better than those that do not. And, those that start planning in later stages do better than when they did none [Hofer and Schendel, 1978]. This suggests that the quality of the strategy process is crucial to how a company performs. The organizational capability for availability engineering and management is integral to that quality.

TO2. The Research and Development Cycle

Research and development is a circle of discovery, learning, and application. Therefore, it leads to change as it eventually triggers and influences the other cycles. This is because it includes the following:

- Research for new knowledge.
- Development of that knowledge into new and advanced products and methods.
- Inclusion of those products and methods in new and advanced plant design and functioning.

Scope of Research and Development

There is often a narrow perception of research and development as focused on products, production processes, and equipment. However, it also includes the following:

- Management and business methods, tools and processes.
- A definition of the capital project and developing associated design processes.
- Market research.

Therefore, the scope of research and development is the quest for improved overall production and business system performance. This is measured by the ability of the organization to use its scarce resources in all aspects of achieving organizational success. This long-run strategic productivity can be improved when research and development explores and develops methods applicable to:

- The efficiency of labor.
- Management effectiveness.
- New technology in the form of ideas, inventions, methods, processes, materials, and computer systems.

The Role of the Availability Discipline

Availability engineering and management is both a contributor and a participant in the research and development cycle. The availability-centered improvement, change and data management functions and systems are an integral part of the research and development cycle. This is because they collect and evaluate data and information from the plant's actual production and support experience.

Living availability design treats the functioning plant as a controlled test laboratory. Thus, it provides processed and evaluated feedback to formal research and development processes.

The living availability design should be regarded as fundamental to the research and development function and its cycle.

The planning cycles for research and development activities will include the details for capitalizing on the availability-centered improvement, change and data management systems and processes.

The plans should not be just limited to the availability performance of current facilities. They must also develop reliability and maintainability experience and insights that can be drawn upon in the development of new technologies.

TO4. The Aggregate Planning Cycle

Definition of Aggregate Planning

The aggregate planning cycle has various names. It is defined here as the cycle of matching productive capacity with forecasted demand. This match incorporates issues of current strategies, resource position, etc. The domain is the owner's company-wide aggregate capacity to produce a product.

The cycle always exists. However, it may be hidden or so closely tied to other management cycles that it is not an obvious or even a named process. However, there is always a process with the purpose of matching productive capacity to market demand and other strategic and constraining issues. An example, is an oil and gas exploration and production organization. Aggregate planning is generally an implicit process within the production department's field development decisions and associated operating budgets.

Aggregate Planning Process

The aggregate planning process is generally as follows:

- Develop forecasts of demand and market share. This includes their cycles which are shaped by business cycles, random events and trends.
- Assess sales orders, sales contracts, and various sales plans for the production requirements they foretell.
- Assess strategic and company-wide issues to be reflected in the aggregate plan.
- Develop strategies for matching plant capacity to demand at the inter-plant or operating company level.
- Allocate production to individual plants on the basis of the low-cost producer.
- Evaluate the effect of the allocation on the various human and material resource pools (i.e., manpower, equipment, and facility resources). This and the four previous steps are iterative.
- Establish the final optimal aggregate plan.

Planning Horizon

The aggregate planning cycle may be monthly. The planning horizon can range up to several years. As previously mentioned, the most basic objective is to match the organization's capacity to meet demand.

This determines the planning horizon. It is important to determine when the owner's system of plants may no longer meet its share of forecasted total market demand. This will be the point at which the plant reaches the outer limits of physically or economically feasible capacity.

The planning time-horizon should be long enough to identify this outer limit. It should allow time to begin the management cycles in the immediate-term to develop the necessary future capacity.

Flexibility for Production System Optimization

There are several implications in the description of aggregate planning. Large operating companies will have multiple production trains, plants and locations for producing the same product. Thus, the objective of aggregate planning at the operating-company level is to optimize production performance and economics for a group of plants.

The production functioning of each plant may be suboptimized. Thus, the capacity to be economically and managerially flexible is a fundamental issue during the commercial production life of a single plant. This is because of the opportunity it provides for formulating advantageous aggregate plans.

Role of the Availability Discipline

Availability engineering and management is fundamental to aggregate planning. This is especially so with regard to the concept of living availability design functions and systems.

Three measures of a successful solution.. There are three fundamental measures of a successful solution to aggregate planning. They are as follows:

- Determining the most profit-effective allocation of available capacity for the required production.
- Producing the product according to specification.
- Producing the product with the most profit-effective approach possible at the plant level.

Availability engineering and management is strongly associated with the first and third goals. The living availability design process is challenged to squeeze productive capacity out of a plant in accordance with these criteria.

Outer limits of feasible performance At some point in the plant's life, no possibilities will remain to hold the cost structure of the availability scheme in the economic valley of Figure 1-10. At a later point, the cost of incremental availability can no longer even be made less than the incremental production value it creates. At an even later point, it will no longer be possible to increase availability performance regardless of spending for field and functional activities.

The alternative is to increase production capacity with capital-based strategies. A possibility is to debottleneck the plant. A greater strategy is a plant addition or a new plant.

The living design systems and availability management functions are basic to these issues since they are concerned with determining a profitable and possible availability scheme. They will also search for the point where there are no longer working capital solutions (operational availability) to availability in productive capacity.

Capacity for dynamic aggregate performance. The living availability design has another role in the aggregate planning cycle. As business cycles oscillate, there will be times that less, rather than more, capacity is desired. One of the most significant and flexible parts of the production cost structure will be the maintenance operations scheme. As availability is reduced so is productive capacity.

The living availability design is an aggregate planning tool for exploring the ramifications of reducing productive capacity. The objective is to determine how to shift the break-even capacity utilization rate downward. In other words, to shift downward the rate the plant must run be to break even. This is done by shifting the current operational availability (A_o) downward.

TO5. The Operating Budget Cycle

Operating Budget and Long-Term Financial Planning Compared

The operating budget cycle is the business process that gives management the image of being concerned only with the short-term perspective. However, its predecessor long-term financial vision and planning process is part of the strategic cycle (TO1).

Like long-range financial planning, the operating budget cycle is usually an annual process. However, it typically has a planning horizon of only one year. It focuses on quarterly or even monthly performance. In essence, the operating budget cycle converts the results of the aggregate plan to various budgets that are related to the owner's short-term financial statements.

The Individual Budgets

The operating budget is a synthesis of individual but related budgets. Each represents a dimension of business performance. The budgets are as follows:

- Revenues.
- Accounts receivable.

- Sales and associated administrative expenses.
- Materials, labor, and overhead budgets for producing the product and, in turn, the cost of the goods-sold budget. The overhead budget will include all or part of the maintenance and support functions.
- An expense budget that includes plant depreciation as a reflection of the plant's design.
- Income statement and balance sheet budgets along with various financial ratios.

The Short-Term of Long-term Financial Planning

The operating budget may appear short-term because the objective of the process is to study and then maximize short-term profits. However, the assumption is that the financial long-term was planned as a different management cycle.

It is also assumed, or hoped, that vital design activities, such as availability engineering, were not omitted from the plant capital project. Omission may be a result of the short-term challenge they present to traditional project tasks and management.

The long-term optimization developed by such processes greatly determines the financial results that can ultimately appear in the operating budget. For example, if the plant is not methodically designed initially to most effectively achieve the specified life cycle availability, its operating budgets will reflect the negative financial consequences. They become conditions that are at least fixed in the short-term. They may possibly even be irreversible.

Consider the consequences of availability engineering during plant design to the operating budget. The plant is designed from the beginning to be able to most profitably produce management's specified life cycle availability. This specification was tied to a typically complex expression of financial, operating and availability performance requirements that were based on complex strategic issues.

Consequently, the period revenues would be increased as they relate to meeting availability goals implicit to the operating budget. Meanwhile, as a result of availability design, the aggregate of direct and indirect maintenance operation and depreciation expenses for the subject accounting periods will be reduced.

If availability engineering was not part of the initial and subsequent design cycles, this omission would be reflected in less attractive income and productivity of assets. Worse, not much could be done to change such an outcome in the short-term or a subsequent longer period of time.

The Role of the Availability Discipline

The issue to be explored in this section is the role of availability engineering and management in the operating budget cycle. The budgeting cycle squeezes maximum short-term income from the owner's plants as they are currently configured. Therefore, it is an iterative process with the aggregate planning (TO4) and the plant production planning (TO6) cycles.

The living availability design is a tool of those iterations with respect to determining what is possible during the operating budget horizon. As such, it is also a tool for generating realistic activity and asset-based budgets for the costs and expenses of availability management and maintenance operations.

The operating budget will also call upon the tools to optimize its system of individual budgets. The various models, especially the resource level calculation models, are utilized in this process as part of the plant maintenance operation planning cycle (TO8).

TO6. The Plant Level Production Planning and Resource Acquisition Cycle

Nature and Scope of the Cycle

There is also a management cycle at the plant level for planning production and its resources. The objective is to convert the aggregate production plan to a shorter-term plan of production. The cycle will be short, whereas, the planning horizon may at least be a year.

The scope of the cycle is generally as follows:

- Develop timing and priorities for various production runs.
- Prepare plans to acquire necessary production resources.
- As the period progresses, develop short-term production plans.
- Frequently compare actual performance to the aggregate plan and operating budget and adjust the plan and resources accordingly.

Role of the Availability Discipline

Availability engineering and management is not a part of this cycle, it is an integrated, parallel cycle. It too is concerned with resource determination and acquisition. Its scope includes the planning and control of plant maintenance and support activities. Therefore, the production and maintenance planning cycles (TO8) are integrated, iterative functions. Each will empower and place constraints on the other.

TO7. The Production Cycle

Scope of the Production Cycle

The production cycle includes the preparation for production, startup, achievement of equilibrium economic production, shut-down for a host of routine and nonroutine reasons, and preparation for the next cycle. It is managed by functions and systems that control capacity, efficiency and flexibility and feed back the results to plant production planning cycles.

Role of the Availability Discipline

The availability scheme is, of course, closely tied to the production cycle just as it is to the production planning cycle. The most direct interface is the many cycles of maintenance and support that are triggered by the current production cycle. There is also the continual process of measuring and recording performance.

TO8. The Maintenance Cycle

As mentioned, availability management through its maintenance cycle will parallel and be integrated with both the production planning (TO6) and production cycles (TO7). The elements of the living availability design will have been involved in some way with most of the previously described management cycles. Those same living design elements will also have a role as tools for supporting the maintenance cycle.

Activities of the Maintenance Cycle

The planning and execution of the maintenance cycle must do the following:

- Convert the production plan to an availability plan that is to be achieved during the planning period.
- Determine the maintenance resources required to serve the availability plan for the period.

- Plan for daily and scheduled maintenance field activities in collaboration with the production operations planning functions.
- Acquire the human and material resources for the planned maintenance operation requirements.
- Execute and control maintenance operations.
- Compare and adjust the actual performance of the maintenance operation activities to meet the operating budget.

In essence, the maintenance cycle is the act of utilizing and fine-tuning the overall plant availability scheme. The tuning serves a current range of planned production levels and scenarios. The availability engineering process will have designed the plant and its availability scheme to provide that capability. Accordingly, it will have left the systems and functions in place that will enable such flexibility to be a management possibility.

On-Line Availability Analysis

The availability model is a valuable tool for on-line maintenance planning. Recall that availability is not a service factor. It is a probability distribution function for expected availability performance.

The plant does not necessarily experience a shutdown or a reduced production rate when an item fails. However, it will always experience a reduction in expected availability. This is because the plant's actual temporary configuration has become progressively different than its designed configuration.

Figure 1-3 showed that the distribution function for availability has shifted to some degree to the left. The distribution function will also take on some less attractive shape with respect to confidence limits. Furthermore, the consequence of a single failure will reflect the current status of all other plant elements as failed, ready, or functioning.

Therefore, short-term decisions in daily planning, such as the placement of a failure in queue for service work, will affect the position of the availability distribution function. In other words, with each failure, inherent availability (A_i) has been reduced. There is a lessor probability of maintaining the specified operational availability (A_o). Thus, the availability model is an on-line tool for assessing the plant's "instantaneous availability." Actions should be scheduled in a response that is consistent with that condition and the desire to avoid an inevitable production reduction or shutdown.

TO9. The Performance Testing and Evaluation Cycle

The only true model on which a plant can be tested and evaluated as a system is the plant itself. Thus, it is crucial that management and the design engineers resist the temptation to regard the commercial production phase as only a matter of product production and its efficiency. The phase must also be respected as the testing and evaluation of a production system. Thus, it is a phase that is actually an adjunct to the research and development cycle (TO2).

Basis of Performance Measurement

Plant performance, as a production system, should be measured against the following:

- The various budgets within the operating budget.
- Input to output ratios with respect to the production process.
- Availability performance against predetermined measures of performance.

Short- and Long-Term Dimensions

The performance testing and evaluation cycle has both short- and long-term dimensions. These dimensions are in many ways a function of where the findings lead when a solution must be developed. The findings will trigger other cycles for their solution. These cycles will be a function of whether performance is to be accepted or improved upon.

Possible cycles of solution will include the following:

- The cycle may lead to the short-term production and maintenance control cycles for fine-tuning performance. This is the case when what is required is found to be still possible.
- The cycle may trigger an aggregate planning cycle. The objective is to adjust its plan to accept what current field activities are now finding to be possible within the existing production facility.

  This may lead to a new allocation of production and resources among plants in the owner's larger production system. It will also appear in some form as a variance in the operating budget. Subsequently, the plant's production capacity may need to be increased by non-capital means for the planning and production periods to follow.

- The findings may trigger the strategic cycle. This may be so when expectations are that the capacity to match demand will no longer be economically attractive or even possible by non-capital means at some identified time in the middle-term future. The result may be strategically driven plant debottlenecking, revamping, or expansion.

TO3. The Product, Production Process, and Plant Development Cycles

The product, process, and plant development cycles are presented here out of its order in the overall production system operations management cycle. It is presented at this time to better show how the other cycles will trigger them.

Triggers of the Development Cycle

Periodically, there will be integrated, iterative activity or sets of subcycles to design a product, its process, and the production system to produce it. The product may be the same. Its composition or production process may be changed. Or the plant may be changed in response to changing production technology requirements such as economics, demand, and market share trends and production inputs. This, in turn, can lead to a change in the product or process.

Almost any of the other cycles can trigger this cycle. Examples are as follows:

- Strategies to achieve or protect a market position, reach productivity goals, design for financial results, etc., can create a need for change.
- Research and development cycles can render the current elements of plant productivity obsolete. They can also discover products and product variations that can be produced if an existing plant is modified in some way.
- Aggregate planning cycles will eventually reveal a specific point in time that the production system will no longer be able to achieve management's strategies.
- Operating budget, production planning and maintenance operation planning cycles may reveal that the plant could achieve much more profit-effective, short-term production results if known new technologies, methods and practices were incorporated in the plant.
- Plant performance evaluation cycles may reveal that capital projects are necessary for the plant to achieve its targets or to improve its performance in specific areas.

Role of the Availability Discipline

The role of availability engineering and management in this cycle is obvious. It will use existing models for each new cycle of design, construction, and startup. However, the cycles will be using progressively better data and information as a result of the availability-centered improvement, change, and data management systems and functions.

Bibliography

Buffa, Elwood S. and Sarin, Rakesh K. *Modern Production Operations Management.* 8th ed. John Wiley & Sons. New York. 1987.

Hofer, Charles W. and Schendel, Dan. *Strategy Formulation: Analytical Concepts.* West Publishing Company. St. Paul, Minn. 1978.

Katz, Daniel and Kahn, Robert, L. *The Social Psychology of Organizations.* 2nd Edition. John Wiley & Sons. New York. 1978.

Tersine, Richard J. *Production Operations Management Concepts.* 2nd ed. Prentice Hall. Englewood Cliffs, N.J. 1984.

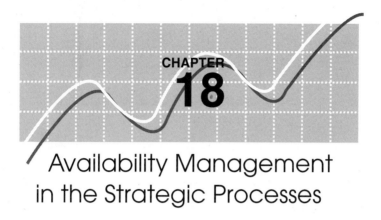

CHAPTER

18

Availability Management in the Strategic Processes

Introduction

The previous chapter introduced nine classic cycles of production system operation and management. A production system was defined as one or more plants connected by management processes or that are part of a value chain. Each was briefly described. A purpose of that introduction was to lay a foundation for the following explorations:

- To identify how the process of availability functions and computer systems, models, and databases formed during the original availability design phases have a role in management throughout a plant's producing life.
- To explore how traditional management is accordingly empowered to produce a better overall business result.

Availability Cycles and Organization

Figure 18-1 shows the management cycles within the plant's producing life. The traditional management cycle processes are above the dark line. They were the subject of Chapter 17. Below the line are the management processes for availability performance. They are integral to the traditional processes. Note that the traditional maintenance operation cycle (TO8) is shown in the overall production cycle.

The availability management processes are as follows:

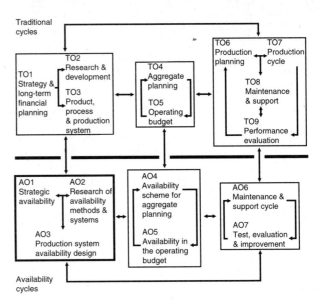

Fig. 18–1 Classic and availability cycles for production system management.

- Analysis of the owner's strategic position with respect to plant availability performance (AO1).
- Research and development of plant availability performance (AO2).
- Availability engineering and management in periodic plant design and modification (AO3).
- Aligning plant availability performance to produce the productive capacity for matching forecasted market demand for its product (AO4).
- Adjusting plant availability performance to maximum short-term financial performance (AO5).
- The maintenance operation cycle (TO8, Chapter 17).
- Availability management in the maintenance operations cycle (AO6).
- Continual testing, evaluation and auditing of the plant availability scheme and performance (AO7).

Availability Processes and Tasks

This and the chapters that follow will collectively describe the overall organizational process to manage availability performance. The process is described generically. There is no attempt to rigorously convert and detail necessary organization functions and structure. Such detail is unique to each plant and its parent organization.

The processes and associated flowcharts are extended to the level of subprocesses and tasks. The reader will note that the task narrative and flowcharts include coded reference to deliverables and their tasks executed during the design phases. Those phases established the elements of availability management as they were first applied. Thus, many are now active as a part of normal availability engineering and management processes throughout the plant's producing life.

Organizational Design

Chapter 13 introduced the ultimate need for the design of the availability organization. Its domain included maintenance operations. A design process was described to fulfill that need. In fact, the ultimate goal of that design is to perform the availability engineering and management processes described in this and the next two chapters. That organization design also includes the maintenance and support cycle (TO8) identified in the previous chapter.

Organization of this Chapter.

The bounded areas of the Figure 18-1 show that this chapter is concerned with the first three management cycles. The common thread is the long-term consequences of their results. They are as follows:

- They approach the plant with respect to achieving some strategic position.
- The time horizon of their consequences is long.
- Their analyses, decisions and ultimate actions affect large parts of the organization.

AO1. Analysis of Plant Availability

Performance for Strategic Position

Corporate and operating company management must make the following strategic decisions:

- A choice of plant, products and product line.
- Choices for service levels in the production and distribution of the plant's products.
- Choices and strategies for competing in a changing product and market life cycle.
- Decisions for the allocation of resources to the chosen strategies.

Nature of Availability Management in the Strategy Process

The strategic cycle for availability will analyze performance with respect to current and potential strategic positions. Each strategy is tested for consistency with the organization's current productive capacity.

There may be a capability gap between the existing and desired position. If so, strategies must be formulated and assessed to bridge it.

Thus, the strategic availability cycle analyzes what is currently possible vis-a-vis each candidate. The fundamental concern is productive capacity. Chapter 1 showed that this is the combined result of the plant's production processes and physical availability to perform. Accordingly, availability management processes will focus on the following:

- What is the current case for availability performance?
- What is possible in terms of incremental availability performance and its economics?
- What is required for each candidate strategy? For example, can the strategies be achieved by realigning resource levels and plant functioning? Or is a capital program required? And if so, what capital investment is required?
- What are the long-term financial implications and variations of the availability requirements? This includes approaches to capacity acquisition in terms of capital versus working assets.
- How must the availability scheme be included in the description of each candidate strategy? These are the details and measures required to fully define the selected strategies. They subsequently guide the implementation of chosen strategies.

Process to Analyze Availability Performance for Strategic Position

The analysis of availability performance for strategic position will draw upon the management elements first created in the design phases. They will be applied to the role of availability management in the overall strategic cycle.

The processes and tasks for analysis are introduced in the following sections. They are flowcharted in Figure 18-2. Its activities are coded as AO1 to depict availability in operations management.

AO1.1. Determine the nature of availability for the proposed strategies. A candidate strategy may change the description of how the plant fits within the owner's larger production system. This is a fundamental issue. The tasks that periodically make this assessment are now subject to review and revision.

The necessary assessment processes were first formalized in the conceptual design phase (Chapter 3). Figure 18-3 is a flowchart of the following process tasks:

AO1.1.1. Strategically define the plant in the owner's total production system. Individual plants are often part of a larger production system. Several plants are often producing the same product. A plant may also be a stage along a value chain of plants.

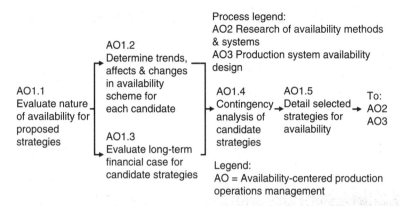

Fig. 18–2 Process to analyze plant availability performance for strategic position (AO1 of Figure 18-1).

The candidate strategies will typically span that system. They will occur as they reflect market and resource environments. Another fundamental dimension is the synergism of owner businesses.

This synergism requires that the fit of the subject plant(s) in each strategy be reviewed (AC1.5). The objective is to establish a policy-level description of the plant's role in the success of each candidate strategy. Plant availability is then described rigorously in the context of the grand production system and the proposed associated strategies (AC2.2 and AC2.3).

AO1.1.2. Define production and availability performance. The previous review and revision process is translated to a definitive description of productive capacity and associated availability performance. The currently defined production scenarios (AC2.4) will be reviewed and possibly revised. The results are extended to define availability performance at multiple production levels in a overall production envelop.

Associated with each level of production is the confidence limits of availability performance. Confidence limits are also a major strategic decision area because of their overall business ramifications. For example, a higher confidence will require a commensurate allocation of corporate financial resources to working capital.

AO1.1.3. Strategic review of the maintenance operation scheme. It is also necessary to review the current maintenance operation concept (AC2). The focus is as follows:

- The integration of the plant's maintenance operations was originally defined for its configuration within the owner's total production sys-

Process legend:
AO1.2 Determine trends, affects & changes in availability scheme for each candidate
AO1.3 Evaluate long-term financial case for candidate strategies

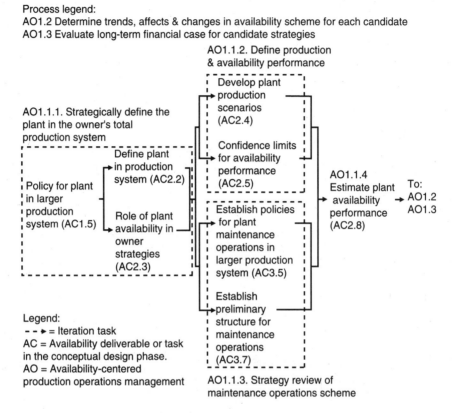

Fig. 18–3 Process tasks to evaluate the nature of availability for candidate strategies (AO1.1 of Figure 18-2).

tem (AC3.5). Thus, it is necessary to determine if each candidate strategy requires that the scheme be revised.

- A candidate strategy may also suggest the need to revise currently established measures of maintenance operation effectiveness (A33.6). For example, a strategic position requiring increased production will change the target utilization rate of fixed support facilities.

AO1.1.4. Estimate plant availability performance. The results of revisiting these conceptual deliverables in the strategic cycle is revised system-level reliability, maintainability, and economic parameters. These measures must be translated to the subsystem level and the calculation of availability performance.

This can be done by revising the preliminary availability model developed in the conceptual design phase (AC2.8). The model is revised to determine the availability performance associated with each candidate strategy.

AO1.2. Determine trends, effects and necessary changes in the availability scheme. The availability performance parameters and elements are defined to a level of resolution necessary for strategy analysis. This is because each strategy causes the achievable and operational availability performance to rest at some point on the profiles shown in Figure 18-4. They will also determine the slope of those profiles.

It is also necessary to identify what must change in the plant. There are four areas of change that will affect the profiles and the plant's position on them. They are hard design; human, material and other support resource levels; maintenance operations; and organizational effectiveness.

Availability trends, affects and changes are studied by reactivating the following availability engineering tasks in the living availability design (Figure 18-5):

AO1.2.1. Review and revise strategic criteria and measures of "best" performance. Some candidate strategies may reveal the need to revise the criteria and associated measures of what is an optimal availability scheme. Thus, the availability design process for identifying and ranking criteria must be reactivated (AB10.1 and AB10.2). Measures must be defined for new criteria (AB10.3). Subsequently, the availability and financial models are modified to incorporate the resulting criteria functions (AB10.4 and AB10.5).

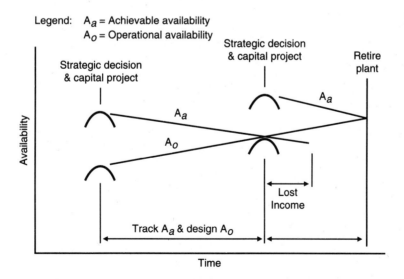

Fig. 18–4 Changing position of achievable and operational availability with time

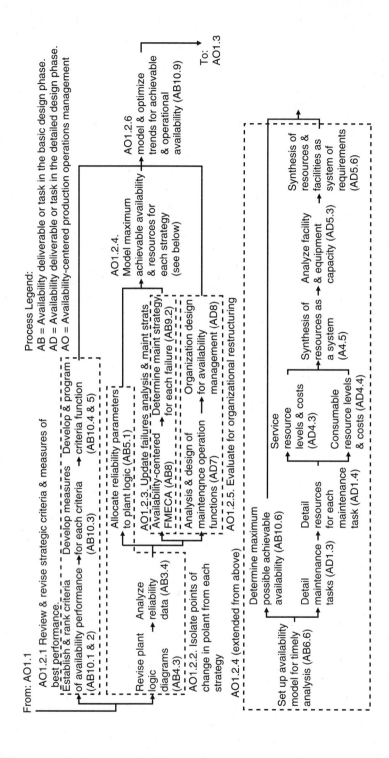

Fig. 18–5 Process tasks to determine trends, effects and changes in the availability scheme for candidate strategies (AO1.2 of Figure 18-2).

336

AO1.2.2. Isolate the points of necessary change in the plant for each strategy. Strategies may somehow change specified plant performance. These must be isolated to points of hard design, functions, resources, and information systems. The points are treated as follows:

- The plant logic diagrams (AB4.3) are reviewed and revised for each strategy. Revision is to add, remove or reconfigure plant elements.

- Candidate strategies can change the operating conditions for some plant items. Availability performance will, in turn, be affected through its reliability component. These affected elements are identified. The logic diagrams are then used to graphically show all plant elements whose reliability performance will be affected by the operating conditions of each strategy.

- The process to analyze reliability parameters of affected plant elements must be then reactivated (AB3.4). Parameters must be reformulated to reflect change over the horizon of the strategy.

- The reliability parameters revised for each candidate strategy are then attached to the logic diagrams (AB5.1).

AO1.2.3. Update failures analysis and associated maintenance strategies. Any necessary change of plant hard design requires a review of failures (AB7) and the determination of a basic maintenance solution for each (AB9.2).

AO1.2.4 Model maximum availability and possible task resources for each strategy. Each strategy must be evaluated for achievable and operational availability. The following availability tools should be applied:

- Install any revised proforma reliability parameters in the availability model (AB6.6). The availability process task for determining maximum possible achievable availability may then be activated (AB10.6). The purpose of this process is to project performance trends over the strategy planning horizon.

- These trends and changes will revise the maintenance task analysis (AD1). There may be new maintenance tasks if the strategy involves physical change. Their steps, time-to-complete, etc. (AD1.3) should be detailed. This, in turn, activates the process of identifying all human, material, and facility resources (AD1.4). Changed operating conditions will also change the frequency of task occurrence (AD1.3.7) and, in turn, affect resource provision levels.

- Operational availability, the lower profile of Figure 18-4, must be projected for each candidate strategy. These are a function of maintenance operation resource levels and organizational effectiveness.

Thus, the models of service and consumable resources and the capacity of maintenance facilities are activated (AD4.3, AD4.4 and AD5.3). The modeling process includes their integration as a system of resources and capacity (AD5.5 and AD5.6).

- Another dimension of strategy is "political." This is the study of the external business and resource environments. Specific to this strategic process is whether the resources are available in the environment. Accordingly, this possibility (fielding analysis, AD4.6) should be reviewed for each strategy. The findings may cause some strategies to be revised or even eliminated from consideration.

AO1.2.5. Evaluate the need for organizational restructuring and development. Functioning associated with availability performance should be also reviewed for necessary organizational change. This is necessary to search for the restructuring requirements associated with any candidate strategy. The focus is on the maintenance operation (AO7) and the overall availability organization (AO8). The design process for both are retraced in search of necessary differences.

AO1.2.6. Model and optimize the trends for achievable and operational availability. The availability modeling system is the tool to aggregate the analysis and findings for the various types of availability performance. It is used to formulate the trends of Figure 18-4 for each candidate strategy. This is done by reactivating the task to model availability performance and its economics (AB10.9).

There are no doubt alternate approaches to availability performance for each strategy. The revised criteria functions are used to determine the best of those schemes for each strategy.

AO1.3. Evaluate long-term financial ramifications of the proposed strategies. The previous processes are extended to formulate long-term financial profiles. These reflect the costs, asset balances, cash flow, and other financial dimensions for each candidate strategy. Long-term financial analysis is empowered by the availability optimization and modeling process (AB10.9). It must ultimately be refined for the selected strategies.

The financial development tasks of the living availability design capability are now activated. This process is shown in the flowcharts in Figure 18-6. The tasks are as follows:

AO1.3.1. Assess the direct life-cycle costs associated with each affected plant element. Direct life-cycle costs include investments in equipment, direct labor, and the materials to maintain them (AB3.5).

The latter two dimensions are extracted from the maintenance task analysis of A01.2. This is available from the estimated direct labor and material costs of each task.

AO1.3.2. Assess the costs and working asset position caused by each strategy. The costs and working asset requirements for organizational functioning and resource requirements must be determined. These include the following:

- The cost of maintenance operations and the overall availability organization (AD2.4, AD6.10, AD7.6, and AD8.9).
- Unit and provision-level costs of human, material, tools and equipment, and facility resources (AD4.3.6, AD4.4.6 and AD5.5).

AO1.3.3. Install cost and asset results in financial model. The direct, functional, and resource costs are installed in the financial models (AB7.8). Direct costs then concurrently flow to and refine the economic parameters allocated to the plant logic diagrams.

AO1.3.4. Model the financial ramifications of each candidate strategy. The financial elements may be modeled in various profiles such as cost and asset balances. The latter includes capital and working assets. Modeling may also profile select financial ratios that are important to the owner's integrated corporate, operating company, and plant strategy analysis processes.

AO1.4. Contingency analysis of the strategy candidates. The candidate strategies have been fully evaluated in terms of their performance and finan-

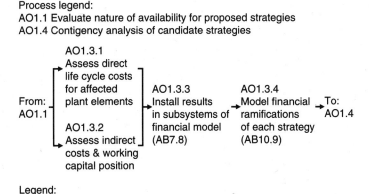

Process legend:
AO1.1 Evaluate nature of availability for proposed strategies
AO1.4 Contigency analysis of candidate strategies

Legend:
AO = Availability-centered production operations management

Fig. 18–6 Process tasks for evaluating long-term financial ramifications of each candidate strategy (AO1.3 of Figure 18-2).

cial ramifications. They must now be subjected to contingency analysis (AB10.10).

Contingency analysis is the inspection of each strategy against unplanned but possible external changes that may substantially impact the chosen strategies.

Any existing contingency scenarios should be reviewed and refined. Some may be dropped. Others may be added. Examples are scenarios for long-term variation in demand, new technology, new regulations, etc.

AO1.5. Detail the selected strategies with respect to availability performance.   Management must ultimately select a strategy or set of strategies. Selection should be in accordance with overall business goals rather than optimal availability performance. This is because availability is just one essential component in the strategic decision. Accordingly, the basic detail of the scheme must be incorporated in the overall description of each strategy.

It is important that this strategic analysis be performed outside of the management of the existing availability scheme. The objective is to devise a preliminary availability scheme. Therefore, strategic availability analysis is approached with only a reasonable degree of rigor.

The availability scheme associated with the set of selected strategies will become the current one. It will then be detailed more thoroughly as the middle- and short-term availability management cycles occur. These cycles are the subject of the next two chapters.

AO2. Research and Development for Availability Performance

Nature and Dispersion of Research and Development

The scope of research and development spans markets analysis, products and productivity. Productivity spans product make-up, production system design, management techniques, computer technology and analysis methods and models. Therefore, the capability for living availability design and its management is a primary area of productivity advancement.

There are two stages of research and development. The first is the development and conversion of basic research to specific knowledge and processes. The second is the application of the knowledge and processes to all aspects of the enterprise. Research and development for availability engineering and management is skewed toward application.

The nature of research and development for availability performance is generally as follows:

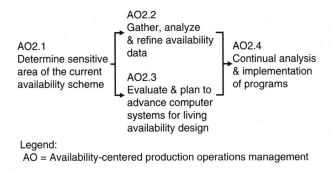

Legend:
AO = Availability-centered production operations management

Fig. 18–7 Process for research and development of availability performance (AO2 of Figure 18-1).

- The refinement of availability analysis methods, models, and data.
- The continual use of the plant as a research model for availability maximization and optimization.
- The conversion of research and methods of the subject plant for use by other plants.

Research and development must not be regarded as an exotic function. It should be seen instead as the basis for many changes, improvements and analytical activities. Its benefits will be dispersed across the many availability and associated maintenance operation processes and functions.

The Process for the Research and Development of Availability Performance

Like all processes, research and development draws upon the availability management elements created in the design phases. The processes and their use of those elements are introduced in the following sections. These processes and their associated tasks are flowcharted in Figure 18-7.

AO2.1. Determine sensitive elements and areas of the availability scheme for refinement. Research and development for availability performance must have focus without losing sight of a bigger picture. In Figure 18-8 it is shown that the focus is formed by the following process for determining elements and areas of refinement in the availability scheme.

AO2.1.1. Inspect the most current strategic analysis and decisions for areas of focus. A review of the most recent strategic process provides great insight to focal areas of advancement. Of the greatest interest are the selected strategies.

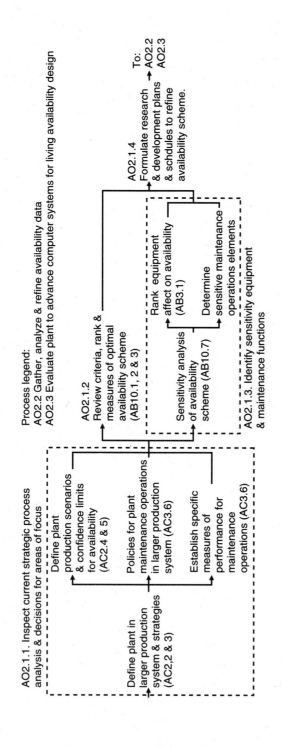

Fig. 18–8 Process tasks to determine sensitive areas of the current availability scheme (AO2.1 of Figure 18-7).

Legend:
AC = Availability deliverable or task in the conceptual design phase.
AB = Availability deliverable or task in the basic design phase.
AO = Availability-centered production operations management

AO2.1.1. Inspect current strategic process analysis & decisions for areas of focus

Define plant production scenarios & confidence limits for availability (AC2.4 & 5)

Policies for plant maintenance operations in larger production system (AC3.6)

Establish specific measures of performance for maintenance operations (AC3.6)

Define plant in larger production system & strategies (AC2,2 & 3)

Process legend:
AO2.2 Gather, analyze & refine availability data
AO2.3 Evaluate plant to advance computer systems for living availability design

AO2.1.2
Review criteria, rank & measures of optimal availability scheme (AB10.1, 2 & 3)

Sensitivity analysis of availability scheme (AB10.7)

Rank equipment affect on availability (AB3.1)

Determine sensitive maintenance operations elements

AO2.1.3. Identify sensitivity equipment & maintenance functions

AO2.1.4
Formulate research & development plans & schdules to refine availability scheme.

To:
AO2.2
AO2.3

342

Important issues to inspect and review are as follows:

- The role of the plant in the owner's larger production system and associated strategies (AC2.2 and AC2.3).
- The plant's production and availability scenarios (AC2.4) and the associated confidence levels of performance (AC2.5).
- Policies for plant maintenance operations in the larger production system (AC3.5).
- The established measures of effectiveness for maintenance operations (AC3.6).

This inspection will reveal areas of focus for research and development. More importantly, they will be driven by the owner's strategic direction. Otherwise, research and development will be conducted at random or driven by the professional interests and political agendas of individuals.

AO2.1.2. Determine the implications of the revised criteria function. The criteria and associated measures of performance were evaluated and possibly revised during the strategic process (AO1.2). The final conclusions should now be inspected as part of the planning of research and development. Of special interest are the criteria, ranking, and measures that have been revised in the most recent cycle (AB10.1, 2 and 3).

AO2.1.3. Identify sensitive equipment and maintenance functions. The availability and financial models are activated for sensitivity analysis (AB10.7). The revealed high and low sensitivities of availability performance to specific plant elements will suggest areas to search for advancement opportunities. The results can be identified along the following dimensions:

- Ranking of equipment assemblies and components for their relative effect on availability performance (AB3.1). This will suggest where new technology and advances in the ability to predict reliability may be most important.
- Ranking of maintenance functions, resource levels, and facility capacities for their relative affect on maintainability. This will suggest where research and development activities may be immediately important to the plant's cost structure and working asset balance.

AO2.1.4. Formulate research and development plans and schedules for refining the availability scheme. The previous findings must be translated to plans and schedules for refining the availability scheme. The objective is to be very specific in the programs to advance plant availability performance and its profitability.

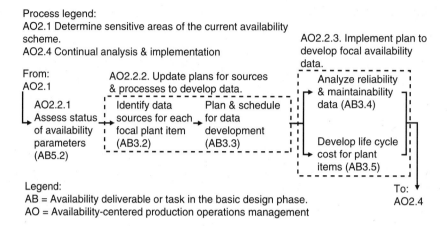

Fig. 18–9 Process tasks for gathering, analyzing, and refining availability data (AO2.2 of Figure 18-7).

AO2.2. Gather, analyze, and refine availability data. A principle of availability engineering and management is that the quality of data is advanced over the plant's lifetime. The driving goal is to continually improve management's ability to predict performance. There are considerable strategic and financial rewards to be gained by striving toward this goal.

Figure 18-9 shows that the tasks for gathering, analyzing, and refining data for the research and development process are as follows:

AO2.2.1. Assess the status of availability parameters. Previous availability development cycles have assessed weaknesses and then strategies to develop availability parameters. Thus, the first activity is to assess their current status (AB5.2) and progress in those strategies (AB5.3).

AO2.2.2. Update the plan for sources and processes for developing data. Review the previously identified sources for data development (AB3.2). The work product will be an updated plan and schedule for data gathering and development (AB3.3).

AO2.2.3. Implement the activities for developing the focal availability data. The activities for gathering and analyzing reliability, maintainability, and economic forecasts should continue (AD3.4 and AD3.5). Implementation will most likely reside with operating company and plant level functions.

Obviously, data development is a persistent activity. Advance planning and management assures a consistent, strategically driven approach. Otherwise, there is a potential for considerable organizational waste. Labor may be misapplied. Meanwhile, a return on costly database systems may never be realized since their content remains of marginal importance.

AO2.3. Evaluate and plan to advance the systems for living availability design. The field of availability engineering and management has been left by history with great room for advancement. One area that has not reached its potential is the devlopment of analytical systems. These tools now exist widely but they have not yet been well developed to function as integrated tools and systems.

Thus, this task is to periodically review and revise the relative stress on advancing various availability systems and their integration. Stress can and should shift with time. One reason is that individual systems will reach a level of advancement in the overall change, improvement, and data management system. Another is for development of systems consistent with the plans to refine sensitive areas and elements of availability performance.

The general process of evaluating and planning the advancement of systems for designing and managing availability performance is as follows (Figure 18-10):

AO2.3.1. Review the current availability systems concept. Review and possibly revise the current concept of availability systems that was developed by the task that produced the current availability concept (AC2.10). Influencing factors are the consequences of new strategies, associated focus for advancing the availability scheme, and the status of the evolving data position.

AO2.3.2. Evaluate the subsystems for data development. The subsystems for data development should be evaluated. This should begin with the subsystem for the continual collection and processing of availability performance data (AB3.7). An extension of the review is the current strategies for data weakness (AB5.3).

The collection systems and availability models may have mechanisms to compensate for these weaknesses. Therefore, advancement may mean change to reflect the evolving quality of data.

AO2.3.3. Evaluate the current ability to model plant availability performance. The capacity to model availability performance should be reviewed for focus. There are the following possibilities:

- The criteria and their measures of performance were previously inspected (AO2.1). The conclusions may have suggested the attractiveness of advancing the current criteria function (AB10.4 and AB10.5). In fact, the development of a rigorous function may present

Process legend:
AO2.1 Determine sensitive areas of the current availability scheme.
AO2.4 Continual analysis & implementation

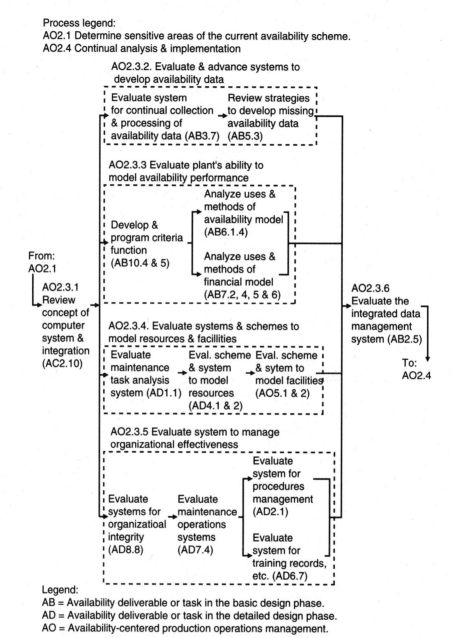

AO2.3.2. Evaluate & advance systems to
develop availability data

| Evaluate system for continual collection & processing of availability data (AB3.7) | Review strategies to develop missing availability data (AB5.3) |

AO2.3.3 Evaluate plant's ability to
model availability performance

Develop & program criteria function (AB10.4 & 5)

Analyze uses & methods of availability model (AB6.1.4)

Analyze uses & methods of financial model (AB7.2, 4, 5 & 6)

From:
AO2.1

AO2.3.1
Review concept of computer system & integration (AC2.10)

AO2.3.6
Evaluate the integrated data management system (AB2.5)

To:
AO2.4

AO2.3.4. Evaluate systems & schemes to
model resources & facilities

Evaluate maintenance task analysis system (AD1.1)

Eval. scheme & system to model resources (AD4.1 & 2)

Eval. scheme & sytem to model facilities (AO5.1 & 2)

AO2.3.5 Evaluate system to manage
organizational effectiveness

Evaluate systems for organizatioal integrity (AD8.8)

Evaluate maintenance operations systems (AD7.4)

Evaluate system for procedures management (AD2.1)

Evaluate system for training records, etc. (AD6.7)

Legend:
AB = Availability deliverable or task in the basic design phase.
AD = Availability deliverable or task in the detailed design phase.
AO = Availability-centered production operations management.

Fig. 18–10 Process tasks for evaluating and planning to advance computer systems for living availability design (AO2.3 of Figure 18-7).

great room for advancement. This is especially so with respect to applied mathematic approaches. Ostrofsky [1977] describes the full possible mathematical development.

- Review the availability model for its potential use in plant life (AB6.1 and AB6.4), and for its current evolution and future possibilities (AB6.2 and AB6.3).
- Evaluate the financial model and its subsystems. This should include the nature and quality of current and future data (AB7.2). It should also include the model structure (AB7.4), life-cycle cost process (AB7.5), and model mechanics.

AO2.3.4. Evaluate the systems for modeling resources and facilities. The systems for modeling resource provision levels and the capacity of maintenance facilities are critical to forecasting performance. Therefore, they are going to be subject to advancement. These include the models for determining service and consumable resource levels (AD4.1) and the throughput capacity of facilities (AD5.1). The maintenance task analysis system (AD1.1) must also be evaluated. This is because it is a core system that capitalizes on the quality of these models. These systems will probably be advanced as an integrated initiative.

AO2.3.5. Evaluate systems for managing organizational effectiveness. The systems for organizational effectiveness in the management of availability performance should be evaluated. This includes the systems for maintenance operations (AD7.4) and overall availability management (AD8.8). Within them are subsystems for procedures and training management (AD2.1 and AD6.7).

AO2.3.6. Evaluate systems for advancement as an integrated total scheme. The above systems should be evaluated individually. However, they must ultimately be evaluated as subsystems in a system (AB2.5). This periodic evaluation has the following issues:

- Are the possible candidate advancements consistent with each other?
- Do they match the sensitive and focal areas of the availability scheme and its performance?
- Are the advancements consistent with the plans for data development?
- Are the priorities, plan, and schedule for development consistent with all other development plans?

AO2.4. Continual analysis and implementation of planned programs. Research and development is ultimately a persistent ongoing analysis of plant availability performance. The plant is the research model since the truest test of availability, reliability and maintainability performance is how the plant actually performs.

The continual analysis processes should be focused by the search for sensitivity in the availability scheme. That search is given focus by corporate, operating company, and plant strategies. The quality of continual analysis is advanced by the enhancement of modeling and data systems. The continual analysis process will also contribute to testing results while implementing planned advancements. The objective in all cases is to better analyze, forecast, and manage availability performance.

AO3. Availability Engineering and Management in Cycles of Plant Development

In Chapter 17 it was explained that a cycle of plant design and construction may occur for reasons in some way related to any of the other cycles. The initiating event may be related to a change of the product, production process, process operation or desired availability performance. These are often driven by strategic issues. In fact, these may all be associated in a single initiating requirement.

Process for Availability Engineering and Management in Cycles of Plant Development

The process for plant design cycles is introduced in the following sections. Figure 18-11 is a flowchart of that process.

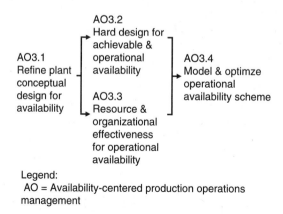

Legend:
 AO = Availability-centered production operations
 management

Fig. 18–11 Process availability engineering and management in cycles of plant management (AO3 of Figure 18-1).

AO3.1.2. Determine system level parameters of performance. The strategic nature of availability performance must be converted to measurable plant performance at the level of plant subsystems. This is to confirm that specified availability performance will be possible. It will establish measures for guiding design decisions.

This begins by studying the evolving plant conceptual design (AC2.1). The plant subsystems should be distinguished with respect to availability performance and their relationship to each other with respect to plant performance (AC2.6). The plant-level availability performance is estimated by running the existing availability model of plant to reflect the new one.

AO3.1.3. Realign the maintenance operation concept. The existing maintenance operation concept must also be refined. This will formulate details and decisions for issues that were not a subject of the strategic cycle. Some possibilities are:

- The organizational entities to be responsible for categories of maintenance tasks (AC3.1).
- Changes and additions to major maintenance support elements (AC3.3).
- Maintenance policies and arbitrary rules (AC3.2).

AO3.1.4. Prepare a final conceptual design document. The availability concept spans both the availability (AC2.11) and maintenance operation (AC3.9) concepts. They are finalized as details, documents, and a unified conceptual statement.

AO3.2. Hard design for achievable and operational availability.   Three dimensions of design occur concurrently. They are as follows:

- Hard design for inherent (A_i) and achievable availability (A_a).
- Maintenance tasks, resources, and facilities.
- Design for organizational effectiveness.

Together, these determine operational availability (A_o). The first dimension is the purpose of this subprocess. The flowchart of Figure 18-13 shows that it will accomplish the following tasks:

AO3.2.1. Revising plant logic diagrams. The plant design is extended to review the plant logic diagrams which must be revised to reflect the new scenario (AB4.3). Plant elements may be added or deleted. They may also be replaced with nonequivalent elements.

AO3.2.2. Determining and acquire data development needs. The revised design will require the development of new availability fore-

Each strategic cycle will have either continued the existing avail scheme or formulated a definitive preliminary design for a new schem now refined and optimized if it constitutes a change of hard design. while, the research and development cycle will have advanced the o ability to design availability performance.

AO3.1. Refine plant conceptual design for availability.

The conc design for availability is possibly revised by each strategic cycle. This c as a necessary process to study and define the nature of availability mance for candidate and ultimately selected strategies.

A design cycle will refine and finalize the conceptual scheme (Fig 12). The focus will be as follows:

AO3.1.1. Refine the defined nature of availability for the se business strategies.

The strategic process assessed the required ability performance for selected strategies (AO1.1). It was con with how the plant and its availability performance is important owner's set of producing plants. It then established productio availability scenarios and the confidence levels for each. This must now be refined for purposes of plant design.

Process legend:
AO3.2 Hard design for achievable & operational availability
AO3.3 Resource & organizational effectiveness for operational availability

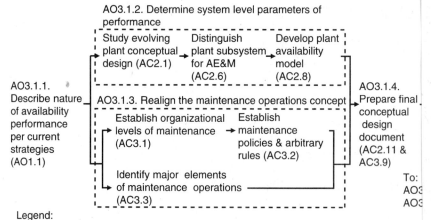

Legend:
AC = Availability deliverable or task in the conceptual design phase.
TD = Traditional deliverable or task in the detailed design phase.
AO = Availability-centered production operations management
AE&M = Availability engineering & management

Fig. 18–12 Process tasks for refining plant conceptual design for availab ity (AO3.1 of Figure 18-11).

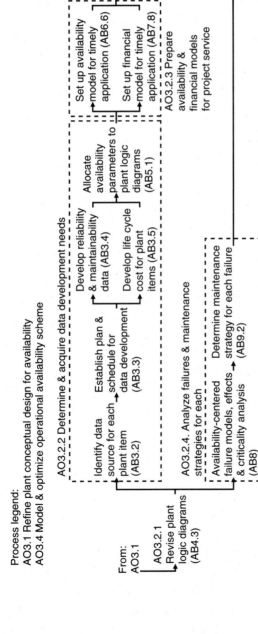

Process legend:
AO3.1 Refine plant conceptual design for availability
AO3.4 Model & optimize operational availability scheme

Fig. 18–13 Process tasks in hard design for achievable and operational availability
(AO3.2 of Figure 18-11)

Legend:
AB = Availability deliverable or task in the basic design phase.
AO = Availability-centered production operations management

casts with respect to reliability, maintainability, and economics. This is especially so for reliability and the accordingly affected economics.

The previous task identified points of physical change. This one is concerned with changed operating conditions. Plant elements will be affected by these changes even though they themselves are not physically changed. The changes are reflected in the expected reliability and economics of the plant elements.

Data as performance forecasts will be both revised and newly developed. The revised case reflects changed conditions. The case of newly developed forecasts reflects changes in plant configuration and components.

The task has the following stages:

1. Review and extend the information for the source-specific challenges to collecting and evaluating data (AB3.2). Because the quality of data will change over time, a new plan and schedule must be formed to gather, analyze, and develop data (AB3.3).
2. Develop reliability, maintainability, and life-cycle cost data (AB3.4/5).
3. Since the work product of the previous steps is availability parameters, the process for allocating parameters to the plant logic diagrams (AB5.1) must now be reactivated.

AO3.2.3. Prepare availability and financial models for project service. The availability and financial models are set up for timely service to plant design (AB6.6 and AB7.8). The first objective is to get them functioning and able to evaluate availability performance as the plant design evolves. They must be programmed to reflect new plant configuration and parameters. Advanced analytical and intermodel integrating mechanisms may also be installed. These should reflect identified sensitivities in availability performance and state-of-the-art possibilities in systems technology.

AO3.2.4. Analyze failures and determine maintenance strategies. The process to analyze failures and a response to each must be reactivated. The objective of this task is to locate the peak on the achievable and operational availability curves of Figure 18-4.

The two stages of this task are as follows:

1. The analysis of failures is studied by the failure modes, effect, and criticality analyses (AB8). The analyses should concentrate on new plant elements. They should also search to find newly created causes and consequences of failures of existing items that are caused by

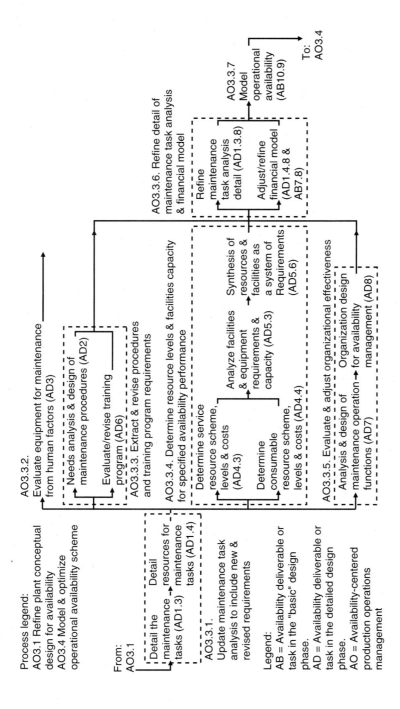

Fig. 18–14 Process tasks for analyzing resources and organizational effectiveness for operational availability (AO3.3 of Figure 18-11).

353

changed plant design. Even though the subject item has not changed, it may still experience changed stress levels.

2. The plant maintenance logic-tree analysis is reactivated (AB9.2) to determine or confirm an appropriate maintenance, operation, or design strategy for each subject significant failure.

AO3.2.5. Model inherent and achievable availability performance. Plant availability performance and its economics must now be modeled (AB10.9). The focus is inherent availability and achievable availability. The objective is to identify the location and trend of the achievable availability profile shown in Figure 18-4.

AO3.3. Analyze resource and organizational effectiveness for operational availability. Operational availability is the bottom line of performance. It reflects the application of resources and organizational effectiveness to complement plant hard design. Hard design defines the plant's capability. Resource levels and organizational effectiveness must be adjusted to produce the current necessary use of that capability.

Accordingly, Figure 18-14 shows that the required tasks are as follows:

AO3.3.1. Update maintenance task analysis to include new and revised requirements. The maintenance task analysis (AD1) is updated to include new maintenance tasks and revise existing ones. The specific areas of change are the detailed description (AD1.3) and detailed resources (AD1.4) of each task.

Many maintenance tasks and, therefore, their steps will at least have a changed frequency of occurrence (AD1.3.7). This is the result of changed plant operating conditions.

AO3.3.2. Evaluate new equipment for human factors. New equipment must be evaluated for maintainability and human factors. The maintenance task detail is used (AD3) to provide the task with focus. The result may be to revise task time, steps, and procedures. The objective is to adjust the detail for actual field conditions.

AO3.3.3. Extract and revise procedures and training programs and requirements. Detail must be extracted from the maintenance task analysis to evaluate the procedures (AD2) and training (AD6) requirements. New procedures and training requirements may be generated. Existing ones may need to be revised. Service levels, organization, costs, and asset requirements are subject to change.

AO3.3.4. Determine resource levels and facilities capacity for the specified performance. The revised maintenance task analysis detail precedes the determination of resource provision requirements, costs, and asset value (AD4.3 and AD4.4). The capacity of existing and new maintenance support facilities (AD5.3) must also be evaluated.

These determinates of performance are integrated in two stages. First, is the integration of service and consumable resources (AD4.5). Second, is the integration of resource levels and capacity of maintenance facilities (AD5.6). This process task is required because all maintenance tasks involve a network of resources and facilities. Their provisioning and capacity are dependent variables in availability performance.

AO3.3.5. Evaluate and adjust organizational effectiveness. Operational availability is partially a function of organizational effectiveness. The focus of this task is as follows:

- The existing maintenance operation organization must be evaluated for revision (AD7) by retracing the existing design. Each finding must be reviewed for its fit with the functioning of the plant once it takes its new form.

- The overall availability organization should be reviewed in the same manner (AD8). This is, in essence, a review of all availability functioning. It should integrate suborganizations for maintenance operations, procedures, training and change, improvement, and data management.

AO3.3.6. Refine the detail of maintenance task and financial models. The detail of the previous processes will ultimately feedback to earlier systems and models. Some inputs to maintenance task analysis are preliminary until this point in plant development. Accordingly, the following must be refined:

- Each affected maintenance task must be detailed as a network of active repair, logistic, and administrative steps and their times-to-complete. The preceding tasks will generate detailed forecasts of logistic and administrative times. The detail of the maintenance task must be refined (AD1.3) accordingly.

- Availability and its maintenance operation costs are both direct and indirect. Asset balances are also an issue. These are determined and detailed by the preceding processes.

Direct costs will pass to the financial model through the maintenance task analysis (AD1.4.8). Indirect costs and asset balances are not directly associated with the availability and associated economic parameters for each plant item. Thus, they flow directly to related subsystems in the financial model (AB7.8). These costs and asset accounts were defined in Part 4 of this book.

AO3.3.7. Model operational availability. The availability model process (AB10.9) must be activated to evaluate operational availability. It was activated previously to determine the achievable availability of the evolving plant design. It is now applied to depict total performance.

AO3.4. Model and optimize the availability scheme. The formulated availability scheme must be optimized. The flowchart (Figure 18-15) shows that the tasks for optimization are as follows:

AO3.4.1. Sensitivity analysis of preliminary inherent, achievable, and operational availability and economics. Sensitivity analysis was part of the research and development of availability methods and performance. Its findings are drawn upon here. They are extended by the search for sensitive and insensitive points in the developing availability scheme (AD10.7). The search should address the following:

• Which elements when varied have the most affect on availability performance and its financial results and

• Which elements when varied have the least consequences for availability performance.

AO3.4.2. Develop variations to the preliminary availability scheme. The findings of the sensitivity analysis have a basic purpose. They reveal where alternatives in the availability scheme may be located and developed for analysis (AB10.8). The search will focus on the high and low ends of sensitivity. A candidate scheme is a permutation of all discovered variations.

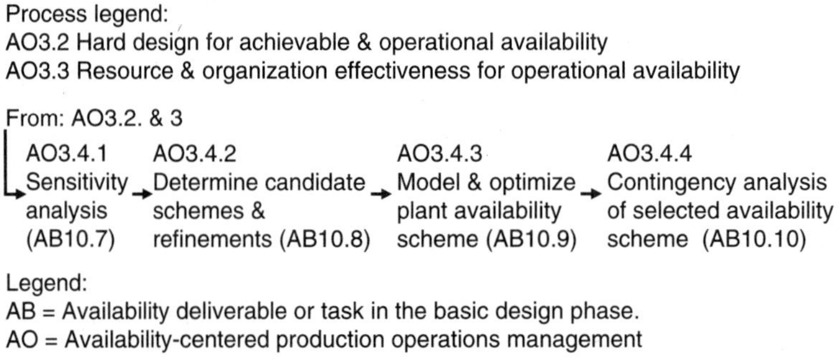

Process legend:
AO3.2 Hard design for achievable & operational availability
AO3.3 Resource & organization effectiveness for operational availability

From: AO3.2. & 3

AO3.4.1	AO3.4.2	AO3.4.3	AO3.4.4
Sensitivity analysis (AB10.7)	Determine candidate schemes & refinements (AB10.8)	Model & optimize plant availability scheme (AB10.9)	Contingency analysis of selected availability scheme (AB10.10)

Legend:
AB = Availability deliverable or task in the basic design phase.
AO = Availability-centered production operations management

Fig. 18–15 Process tasks for modeling and optimizing the availability scheme (AO3.4 of Figure 18-11).

AO3.4.3. Model to determine best availability scheme. The candidate schemes are analyzed by modeling. The objective is to find the best one (AB10.9) among many feasible candidates. This determination is made against a predetermined criteria function. The function should be reviewed and possibly revised as a part of each strategic cycle.

AO3.4.4. Contingency analysis of chosen availability scheme. The findings of the optimization are subject to contingency analysis (AB10.10). Previous scenarios are reviewed for relevance. They may be reformulated to reflect events, history, and new insights.

Summary

The three long-term oriented cycles are interrelated. This is apparent by their description. They jointly span the entire availability design process described for the plant's original design phase.

However, there are differences with the initial design phases. One is that the strategic cycles draw upon existing availability engineering and management infrastructure. They exist as a work product of the original design phase. Thus, they enable the strategic cycles to achieve a heretofore unobtainable level of competence.

Another difference is that the focus of availability engineering and management is on "what is new" and "what is affected." Thus, the cycles of this chapter do not require a full plant design. Furthermore, the development of each new scheme will draw upon an increasingly improved ability to design, analyze and optimize availability performance.

This same case can be seen in the chapters that follow. The scheme of the long-term cycles will be further refined to serve shorter-term performance needs and optimization.

Bibliography

Buffa, Elwood S. and Sarin, Rakesh K. *Modern Production Operations Management.* 8th ed. John Wiley & Sons. New York. 1987.

Hofer, Charles W. and Schendel, Dan. *Strategy Formulation: Analytical Concepts.* West Publishing Company. St. Paul, Minn. 1978.

Ostrofsky, Benjamin. Design, *Planning and Development Methodology.* Prentice Hall. Englewood Cliffs, N.J. 1977.

Tersine, Richard J. Production *Operations Management Concepts.* 2nd ed. Prentice Hall. Englewood Cliffs, N.J. 1984.

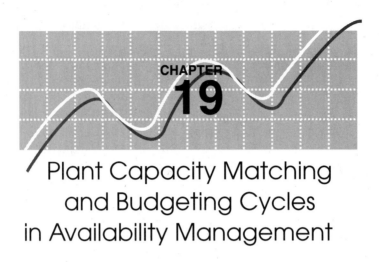

19

Plant Capacity Matching and Budgeting Cycles in Availability Management

Introduction

Chapter 18 described the long-term cycles of availability performance. These cycles are critical to management's ability to undertake the larger equivalent classic cycles they are part of. The bridge between the strategic and short-term production cycles are the subject of this chapter. Figure 19-1 shows that the subject processes for availability management are as follows:

- Optimizing availability as part of matching productive capacity to the owner's forecasted market demand (AO4).
- Adjusting availability to maximize short-term financial performance (AO5).

AO4. Adjusting Availability to Match Plant Productive Capacity to Forecasted Market Demand

An availability engineering and management cycle is undertaken to achieve the plant-level availability that is consistent with the aggregate production planning cycle (TO4, Chapter 17).The objective is to make optimal

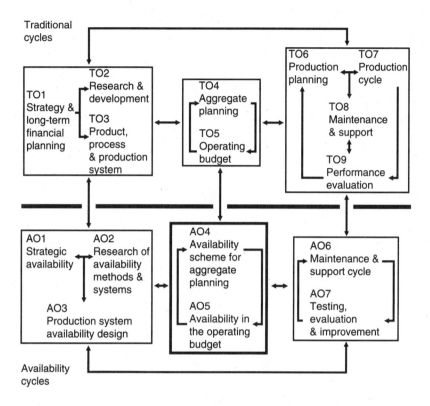

Fig. 19–1 Cycle to plan and budget utilization of plant capacity for meeting market demand.

use of existing production facilities to meet forecasted demand trends, business cycles, and random fluctuations.

Goals of the Availability Discipline in Aggregate Planning

The identification of the optimal production scheme may lead to requirements to either increase or decrease plant availability performance. This is in association with an adjustment in the production level. Thus, the associated availability engineering and management cycle is directed to that end. It will both analyze and plan availability through the aggregate planning horizon. The points of concern may be any of the following:

- To determine the consequences of the assigned production to the plant as a system.
- To determine which aspects of the maintenance operation scheme can be realigned in the planning horizon.

- To determine and detail the adjusted availability scheme.
- To determine when during the aggregate plan horizon it will no longer be economically feasible to attempt such adjustments and capital-based approaches will be necessary to fulfill the aggregate plan.

This point is the point of convergence in Figure 18-4. To the left of that point, management will lose income if it attempts to achieve its availability needs through working capital approaches.

Process to Adjust Availability to Serve Forecasted Market Demand

The process for aggregate planning is introduced in the following sections. They are flowcharted in Figure 19-2.

AO4.1. Convert assigned production to availability descriptions. The first availability management process is to realign the currently standing availability scheme. The tasks are shown in Figure 19-3. They are as follows:

AO4.1.1. Define the plant within the larger production system. The aggregate planning process assigns production requirements across the owner's "system" of plants. Thus, the definition of the plant within that larger system is revisited (AC2.2) and updated accordingly.

AO4.1.2. Refine the plant production and availability scenarios. The latest edition of the plant production scenario (AC2.4) is reviewed and refined. It will be a holdover from the most recent strategic or aggregate planning cycle.

The confidence limits are reviewed as part of each aggregate planning cycle (AC2.5). They are also reviewed with respect to ongoing research and development (AO2). It may have improved the capacity to forecast

AO4.1	AO4.2	AO4.3	AO4.4
Convert assigned production to availability description	Determine affected elements & subsystems	Forecast reliability parameters as a function of operating conditions	Forecast resources & working assets

Legend:
AO = Availability-centered production operations management

Fig. 19–2 Process to adjust availability to match plant productive capacity to forecasted market demand (AO4 of Figure 19-1).

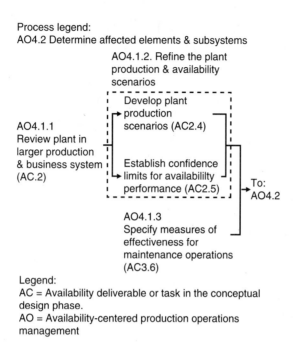

Process legend:
AO4.2 Determine affected elements & subsystems

Fig. 19–3 Process tasks to convert assigned production to availability description (AO4.1 of Figure 19-2).

availability performance within narrower confidence limits. The objective is to reap the rewards of such advancement.

AO4.1.3. Review the established measures of effectiveness for maintenance operations. The performance of maintenance operation is guided and evaluated against predetermined measures of effectiveness (AC3.6).

This task is to review these measures for their fit to the current aggregate plan and possibly refine them. This is the case as experience with the plant is converted and managed against ever finer measures. The research and development cycles may have improved the ability to establish and control against these measures.

These process tasks act to define the production requirements over the planning horizon. Thus, their results are to define plant level performance in measurable terms.

AO4.2. Identify affected subsystems and elements. The results of the previous process may be extended to identify which plant subsystems and their elements are affected by the allocated production. The process tasks are flowcharted by Figure 19-4. They are as follows:

Process legend:
AO4.1 Convert assigned production to availability description
AO4.3 Forecast reliability parameters as a function of operating conditions.

From AO4.1 →	AO4.2.1 Distinguish plant subsystem of availability performance (AC2.6) →	AO4.2.2 Adjust the parameters of the availability model (AC2.8) →	AO4.2.3 Develop plant logic diagrams (AB4.3) →	To: AO4.3

Legend:
AC = Availability deliverable or task in the conceptual design phase.
AB = Availability deliverable or task in the basic design phase.
AO = Availability-centered production operations management

Fig. 19–4 Process tasks to identify affected subsystems and elements (AO4.2 of Figure 19-2).

AO4.2.1. Determine subsystem level availability performance for assigned production. The plant subsystems have been distinguished as a normal task in each plant design cycle (AC2.6). These must be reevaluated for their response to the current aggregate plan.

AO4.2.2. Adjust the parameters of the availability model. The plant availability model (AC2.8) may be adjusted to depict necessary performance during the aggregate planning horizon. The model shows the reliability, maintainability, and economic parameters allocated to the plant subsystems. These should be developed to define proforma performance over the planning period.

AO4.2.3. Identify affected plant elements on the plant logic diagrams. The plant logic diagrams (AB4.3) can be used as a checkoff and logic tool to identify very specifically what plant items are affected by new operating conditions. These conditions are the result of production assigned to the plant.

AO4.3. Forecast reliability parameters as a function of operating conditions. The subsystem level availability requirements of production assigned to the plant have now been isolated to specific plant elements. Their response to operating conditions will determine plant-level availability performance. Therefore, this process task evaluates and formulates reliability parameters as follows (Figure 19-5):

AO4.3.1. Plan and schedule the development of data. Prior to developing reliability parameters, it is important to plan (AB3.3). One objective is

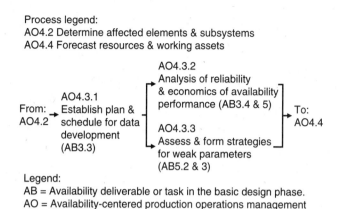

Process legend:
AO4.2 Determine affected elements & subsystems
AO4.4 Forecast resources & working assets

Fig. 19–5 Process tasks to forecast reliability parameters as a function of operating conditions (AO4.3 of Figure 19-2).

to determine sources, the nature of the data, and challenges involved in its acquisition and development.

AO4.3.2. Forecast reliability and associated economic parameters for each of the affected elements. The reliability and economic parameters are determined for each affected element (AB3.4 and AB3.5). They are forecasted as a trend over the aggregate planning horizon.

The life-cycle data management system will have continuously collected and processed data for applications such as these. The system for data acquisition was initially defined in the basic design (AB3.7) phase. It is now periodically updated to identify weak points in important data, a scheme for rectifying them over time, and the associated system mechanics.

AO4.3.3. Review long-term strategies for data development. It has been mentioned that data for reliability development will have weaknesses. These are progressively reduced with time. Therefore, weaknesses in parameter forecasts (AB5.2) should be periodically assessed by virtue of this cycle. The objective is to review and refine strategies for their advancement (AB5.3). This may lead to changes in the system for continual collection and processing of data (AB3.7).

AO4.4. Forecast resources and working assets. The next process is to develop a profile of forecasted achievable and operational availability. The primary objective is to study the latter. This is because it is a function of resource levels, the capacity of the maintenance facilities, and organizational effectiveness. However, the alignment of resources is the focus of the aggregate planning cycle.

The tasks are flowcharted in Figure 19-6 as follows:

AO4.4.1. Allocate reliability and maintainability parameters to logic diagrams. The revised reliability and maintainability parameters are allocated to the plant logic diagrams (AB5.1). They and their associated economic parameters flow to the availability and financial models and the maintenance task analysis detail.

AO4.4.2. Determine material and human resources. As mentioned, the aggregate plan is resource oriented. It is concerned with hard design only as a predecessor to the computation of operational availability (A_o). Thus, the human and material resource (AD4.3 and AD4.4) and facility capacity models (AD5.3) are activated. Their results are integrated as a system of resources and service capacity (AD4.5 and AD5.6).

AO4.4.3. Refine resource requirements in maintenance operation detail and tasks. The results of the resource and facilities analysis flow in two directions. This is a function of whether the subject elements of operational availability are directly or indirectly related to maintenance tasks:

- Direct resource time and costs flow to the detail of maintenance task analysis (AD1.3 and AD1.4).

- The indirect expenses and working asset balances flow to the various subsystems of the financial model (AB7.8).

AO4.4.4. Model availability performance and confidence limits for production allocated to the plant. The modeling process for availability is activated (AB10.9). The result is a proforma depiction of achievable and operational availability. Proforma usage, asset balances and management costs for resources are also projected. Consequently, achievable availability has been translated to resources plans required to achieve the production assigned to the plant.

Figure 18-4 showed that achievable and operational availability will converge at some time in the future. Greater availability for creating productive capacity can then only come from plant redesign. To increase resources and organization effectiveness is an attempt to fill a glass beyond its rim. Plant income will be reduced significantly whenever this point occurs but cannot be identified.

Thus, availability modeling for aggregate planning must identify that point. The aggregate planning horizon should be long enough for mak-

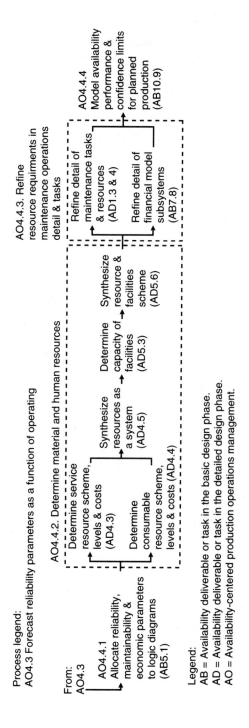

Process legend:
AO4.3 Forecast reliability parameters as a function of operating

AO4.4.3. Refine
resource requirments in
maintenance operations
detail & tasks

From:
AO4.3

AO4.4.1
Allocate reliability,
maintainability &
economic parameters
to logic diagrams
(AB5.1)

AO4.4.2. Determine material and human resources

Determine service
resource scheme,
levels & costs
(AD4.3)

Determine
consumable
resource scheme,
levels & costs (AD4.4)

Synthesize
resources as
a system
(AD4.5)

Determine
capacity of
facilities
(AD5.3)

Synthesize
resource &
facilities
scheme
(AD5.6)

Refine detail of
maintenance tasks
& resources
(AD1.3 & 4)

Refine detail of
financial model
subsystems
(AB7.8)

AO4.4.4
Model availability
performance &
confidence limits
for planned
production
(AB10.9)

Legend:
AB = Availability deliverable or task in the basic design phase.
AD = Availability deliverable or task in the detailed design phase.
AO = Availability-centered production operations management.

Fig. 19-6 Process tasks to forecast resources and working assets (AO4.4 of Figure 19-2).

ing an orderly response to that requirement. This includes the time to proceed through the owner's decision-making and implementation processes for capital projects.

AO5. Adjusting Availability to Maximum Short-Term Financial Performance

Achieving the desired short-term productive capacity is not just a matter of physical performance. More importantly it is a matter of income and productivity of working assets. This is partly a consequence of having optimally allocated the needed production capacity across the owner's system of plants. It is also a fundamental purpose of the operating budget cycle (TO5, Chapter 17).

Availability management is fundamental to this purpose. This is because other than production materials, maintenance operation resources are the most flexible and significant short-term aspect of the plant cost and asset structure. This is equally true at the operating-company level.

Process for Adjust Availability to Maximum Short-Term Financial Performance

The business processes for availability management in the operating budget are introduced in the next sections. They are flowcharted in Figure 19-7.

AO5.1. Analyze the sensitivity of availability to resources. An objective of the operating budget process is to maximize short-term income. Accordingly, the availability systems are used to explore and search out maximally profitable resource schemes. In other words, to find the best scheme to produce the now specified availability performance.

This begins with sensitivity analysis focused on resources. There are various measures that such analysis will test against. Possibilities are as follows:

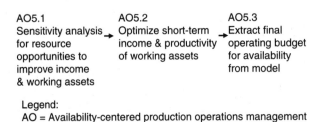

Fig. 19–7 Process tasks to adjust availability to maximum short-term financial performance (AO5 of Figure 19-1).

Process legend:
AO5.2 Optimize short-term income & productivity of working assets

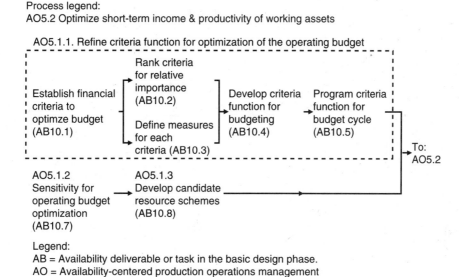

Legend:
AB = Availability deliverable or task in the basic design phase.
AO = Availability-centered production operations management

Fig. 19–8 Process tasks for sensitivity analysis for resource opportunities to improve costs and productivity of working assets (AO5.1 of Figure 19-7).

- Least total expenses for resources in the planning period.
- Least working capital requirements to back up resource levels such as payroll and inventory.
- Maximum income as a combination of the created availability performance versus the expense to achieve it.
- Maximum return on investment to reflect the optimal balance between income and productivity of working assets.

There are many possibilities. They will reflect current business cycles, the owner's desired and necessary financial position, etc.

The operating budget process will fine-tune the availability scheme developed during the aggregate cycle. The flowchart of Figure 19-8 shows that the sensitivity process will activate the living availability design as follows.

AO5.1.1. Formulate a criteria function specifically for optimization of the operating budget. Aspects of the criteria function have been developed and revised by the strategy cycle. Criteria must now be established and ranked specifically for optimizing the operating budget (AB10.1 and AB10.2). The measures for each are determined (AB10.3) and formed into criteria functions (AB10.4). If the resulting criteria

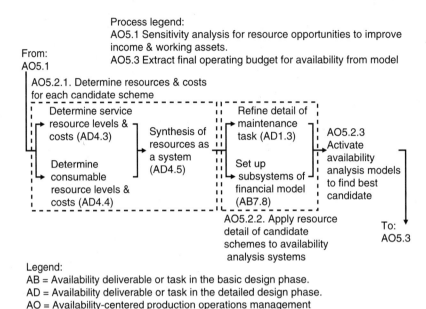

Process legend:
AO5.1 Sensitivity analysis for resource opportunities to improve income & working assets.
AO5.3 Extract final operating budget for availability from model

From:
AO5.1

AO5.2.1. Determine resources & costs for each candidate scheme

Determine service resource levels & costs (AD4.3)

Determine consumable resource levels & costs (AD4.4)

Synthesis of resources as a system (AD4.5)

Refine detail of maintenance task (AD1.3)

Set up subsystems of financial model (AB7.8)

AO5.2.3 Activate availability analysis models to find best candidate

AO5.2.2. Apply resource detail of candidate schemes to availability analysis systems

To:
AO5.3

Legend:
AB = Availability deliverable or task in the basic design phase.
AD = Availability deliverable or task in the detailed design phase.
AO = Availability-centered production operations management

Fig. 19–9 Process tasks to optimize short-term income and productivity of working assets (AO5.2 of Figure 19-7).

function is different than that developed for the last budget cycle, it is programmed into the availability and financial models (AB10.5).

AO5.1.2. Analyze sensitivity for operating budget optimization. The models will be used for sensitivity analysis (AB10.7). The difference with previous cycles of optimization is the focus on the cost structure elements that are flexible in the short-term.

AO5.1.3. Formulate candidate resource schemes. Alternative resource schemes are formed at the elements of high and low sensitivity for availability economics and performance. Each permutation of variations is a candidate scheme.

AO5.2. Optimize short-term income and working assets. The criteria function is now applied to the candidate schemes. The objective is to select the maximally profitable candidate. This will involve the following tasks (Figure 19-9):

AO5.2.1. Determine resources and their cost for each candidate scheme. Consumable and service resources are optimized as a system for each candidate. This will activate the two resource models (AD4.3, and AD4.5). The primary flexible elements are human resources and spare/repair parts inventories.

AO5.2.2. Apply the resource detail of each candidate to analysis systems. The determined resource levels will be applied to the following availability systems prior to the process of selecting the best scheme.

- The detail of maintenance tasks for time to undertake the tasks (AD1.3.8).

- The detail of the financial model subsystems (AB7.8).

AO5.2.3. Activate availability analyses models to find the best candidate. The availability optimization and modeling process must be activated (AB10.9) to analyze each candidate against the operating budget criteria function. The best schedule is selected.

AO5.3. Extract the final operating budget for availability performance. The last process is to harvest the results of the previous two. They are captured in the financial model (AB7.8). It is now only a matter of extracting them as an operating budget.

The financial model has been formatted to match the owner's financial statements and provide performance measures and ratios for control. Thus, it has been designed to be a normal tool in the operating budget cycle.

Summary

The nature of the two processes of this chapter is to refine the work products of the strategic or long-term cycles. In fact, the long and mid-term cycles draw upon common processes.

The only real difference is the level of resolution as a function of goals for each cycle. In turn, the production cycles will operate and control against the results of the mid-term cycles. The production processes for availability performance are the subject of the next chapter.

Bibliography

Buffa, Elwood S. and Sarin, Rakesh K. *Modern Production Operations Management.* 8th ed. John Wiley & Sons. New York. 1987.

Ostrofsky, Benjamin. *Design, Planning and Development Methodology.* Prentice Hall. Englewood Cliffs, N.J. 1977.

Tersine, Richard J. *Production Operations Management Concepts.* 2nd ed. Prentice Hall. Englewood Cliffs, N.J. 1984.

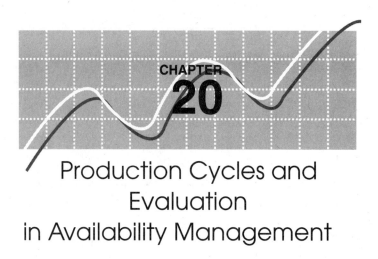

Production Cycles and Evaluation in Availability Management

Introduction

This chapter is concerned with the cycles of short-term production and performance evaluation. The availability management processes direct, support, and measure the short and very short classic maintenance operation cycles. As shown in Figure 20-1, the availability management cycles of this chapter are as follows:

- Maintenance operation cycle planning and functioning (AO6).
- Testing, evaluatiing, and auditing of the availability scheme as a system (AO7).

AO6. Maintenance Operation Cycle Planning and Functioning

Chapter 17 explained that plant production cycles are planned, resources are acquired, and the results are implemented under control. The availability scheme has parallel short-term planning, resource acquisition, implementation and control functions.

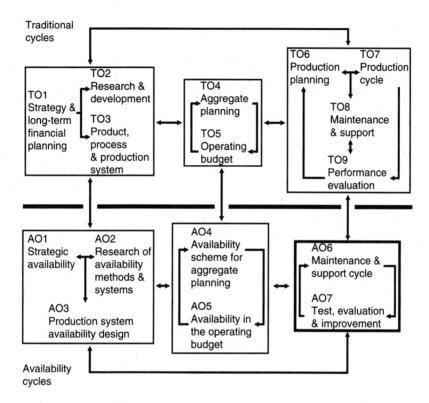

Fig. 20–1   Cycles of production and evaluation in availability management.

Maintenance Cycle Function

The functions and cycles of maintenance were introduced in Chapter 13 (AD5). The maintenance operation cycle was introduced in Chapter 17 (TO8). Both were described in the context of the short-term production cycle.

The activities associated with these classic functions will include the following:

- Near-term and daily planning of maintenance activities.
- Planning and resource acquisition for major maintenance events.
- Execution, control, and quality assurance of field activities.
- Control and functioning of maintenance support functions and their management tools.

Availability Management Process for Maintenance Operations Cycle Planning and Control

Management in the maintenance cycle with respect to availability performance will draw upon the elements created in the design phases. The resulting processes are introduced in the following sections and flowcharted in Figure 20-2.

AO6.1. Dynamic Daily planning with inherent availability (A$_i$). Fundamental to the maintenance operation cycle is daily and very short term planning. This is also basic to the traditional maintenance operation cycle.

A dimension of performance is inherent availability. This was previously defined as expected availability between planned shutdowns for preventative maintenance.

Chapter 7 explained that expected availability is reduced with each failure. However, availability is not defined as lost until production is forced to a lower level. The probability of continued availability at a specific level decreases until it is finally lost.

Thus, change is continuously taking place in the plant. However, it cannot be fully sensed until there is finally a tangible reduction or loss of plant output.

The availability model is a tool for managing the dynamic nature of availability performance. It is used on-line to model the consequences of each new failure in terms of expected availability performance. This determination will also reflect the consequences of existing failures as each new one is added.

The planner can formulate a tactical response to existing and new failures with respect to the hidden changes they create. The model of inherent

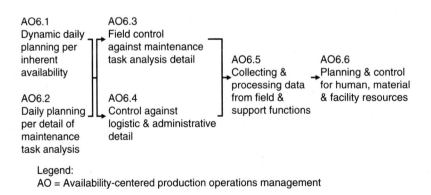

Legend:
AO = Availability-centered production operations management

Fig. 20–2 Process for maintenance operation cycle planning and control (AO6 of Figure 20-1).

availability is also used to devise an optimal path "out" of the real-time configuration of failures in the plant.

The ramifications of real-time availability analysis are immense. For example, can the plant still achieve an average production level but with fewer and lower performance peaks? The plant cost structure would be substantially reduced. There will be an even much greater percentage of change in profits. Another possibility is that resource levels can be maintained within narrower confidence limits. This is simply the results of repeatedly making better tactical decisions. The reader is left to their imagination to ponder other implications.

AO6.2. Daily planning with the detail of the maintenance task analysis. The maintenance task analysis has detailed the steps, times, resources etc., of all significant failures (AD1.3 and AD1.4). This is a living process. Thus, the captured detail is the current best detail and forecast of all requirements and information.

The maintenance operation processes will draw upon the detail to plan and then prepare documentation for each task. This process is automated as part of the availability scheme. This is made possible by the processes and databases of the various maintenance operation computer systems.

AO6.3. Field control against maintenance task analysis detail. A core availability performance process is to control field activity against the maintenance task analysis detail. The purposes are:

- To assure the plant will perform as planned.
- To determine when forecasted availability performance is not realistic.

The control of performance against detail is a critical availability management process. Organizational effectiveness in this role is crucial to progressively improving the owner's business results. This is one of the reasons that the design and ongoing assessment of the overall availability scheme is a living process.

A number of maintenance operation roles and elements are critical to these control process. They are as follows:

- Maintenance operation functions assure the integrity of the task environment, whether or not the maintenance work performed is according to intended standards and procedures and quality of collected data.
- Subsystems of the data management system were designed to collect and process data. The products will flow to data bases and automated analytical processes.
- Maintenance operation functions will determine if the field results belie the planned performance that was expected by virtue of availability

analysis. Alternately, the same functions will determine if the subject field activity requires remediation.

AO6.4. Maintenance support control against logistic and administrative detail. There are a considerable number of logistic and administrative functions in the maintenance cycle and functioning. Like field activities, they too must be controlled against the detail of the current availability scheme.

There are multiple sources of control data and information for logistic and administration processes. They are as follows:

- The process and linkage diagrams of maintenance operations define the flow of logistic and administrative tasks (AD7.1).
- Time to accomplish each logistic and administrative task has been estimated (AD7.7).
- Each maintenance task is detailed as the network of active repair, logistic, and administrative steps (AD1.3.6). Time to complete is attached to those steps (AD1.3.8).

AO6.5. Collecting and processing data from the field and support activities. Availability management processes for planning and controlling against detail lead to another process. That is the process of collecting and processing data from field and support activities.

Data is collected into various databases. Many are associated with maintenance operations. The availability-centered data management system (AB2.5) will have defined a scheme to gather, process, integrate, and extract data from them. This scheme will have been based on the following:

- How is maintenance performance to be measured?
- What data products are required for management to monitor and control against those measurements?
- What processes, calculations, and products are required to produce them?
- What data elements are required?

AO6.6. Planning and control for human, material and facility resources. Logistic time is a function of planning and control of human, material, and facility resources. Thus, there are also availability management processes associated with resources.

Their basic objectives are to:

- Validate and refine forecasts of usage.
- Determine when the various resource control schemes must be reviewed.
- Maintain active control of resource levels.

These processes depend on the preceding one to manage data. Validating task logistic and administrative detail, in turn, enables resource levels to be refined. The capacity to function within narrower confidence limits will also evolve. The quality of controlled, collected and processed data is crucial to such improvement.

AO7. Test, Evaluation and Audit of the Availability Scheme as a System

No matter how well designed, the availability scheme is still based only on calculations and assumed operating conditions. This is the case for all dimensions of plant design and performance.

Thus, it is necessary that the availability scheme be subjected to cycles of testing and evaluation. The purpose of these cycles is to

- Measure actual performance against that promised by the designed availability scheme.
- Determine if short-term plant performance is meeting its targets.
- Improve management's capacity to predict performance.
- Perform audit processes.

The availability-centered improvement, change, and data management function and systems are an integral part of the cycle. They will gather and evaluate data and information with the various availability engineering databases and models. The availability management functions will often spearhead the use of that information and data for improvement, change, and auditing.

Availability Management Process for Testing, Evaluation and Auditing of the Availability Scheme as a System

The availability management processes for analysis are introduced in the next sections. They are flowcharted in Figure 20-3.

AO7.1. Evaluate actual availability against designed performance. Evaluating forecasted versus actual performances is not particularly meaningful at an aggregate level. However, evaluation of availability performance can be thorough and element-specific because of the availability analysis models and systems. The following determinations can be made:

- Did plant level performance match specified performance?
- Where within the many plant elements was there a variation between actual and expected performance?

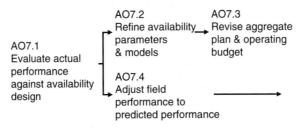

Legend:
AO = Availability-centered production operations management

Fig. 20–3 Process for test, evaluation and audit of the availability scheme as a system (AO7 of Figure 20-1).

The second case is the most important even if there is little variation in plant level performance. The search will be for variations in active repair, logistic, and administrative elements of each maintenance task. Reliability evaluation will be the focus of activities to test middle and longer term performance.

The capacity to test against measures for individual elements is critical for continuously improving business results for the following reasons:

- Good aggregate performance can still hide poor performance. For example, one performance item or activity may do well. Meanwhile, poor performance for another may be offset by that positive result.
- There will be opportunities to advance the performance of elements to which plant availability performance is sensitive.
- There will be concurrent opportunities to capitalize on the high performance success of key elements.
- The combined effect will allow the availability scheme to be adjusted downward as achievable availability is shifted upward and operational availability remains as specified. Thus, the cost and asset structure of the currently necessary operational availability will be reduced.

AO7.2. Refine prediction of availability parameters. The results of the previous activity are used to refine the reliability, maintainability, and economic parameters. They ultimately flow to the availability and financial models. They also reflect revisions through the resource, facility, and maintenance tasks analysis models and systems.

AO7.3. Revise the aggregate plan and operating budget. Each cycle of evaluation and auditing will advance management's ability to predict performance. Each such advancement should be immediately capitalized upon.

The most immediate possibility is to refine the operating budget. This will advance management's effectiveness in short-term control toward a higher maximum result.

Another immediate possibility is to refine the current aggregate plan. The findings will revise management's current perception of the relative positions of achievable and operational availability. The ability to refine the aggregate plan will affect asset balances to achieve the required production. It may also cause the optimization across the production system plants to be advanced.

However, the payoff is not better budgets and plans. It is management's ability to identify and then execute better business results.

AO7.4. Adjust field performance to match predicted performance.   The key to refining performance is to adjust field and support performance to forecasted performance because such elements are subjected to brief or intensive analysis or auditing processes. This may include plant process operations and control. The specified elements are then subject to remedial actions identified by analysis and auditing. Success is measured against previously established reliability, maintainability and economic parameters for each element.

Commercial Production without Availability Engineering and Management

All through this book, the value of availability engineering and management has been explicitly and implicitly identified. However, it may also be revealing to describe the plant without it.

Unknown Initial Condition

In the beginning, the plant can only be expected to achieve some adequate level of availability performance. Only by the longest odds can management expect that a specified availability may be possible, and by even longer odds that it is the most profitable scheme. Worse, management will not know the plant's inherent, achievable, and operational availability. Nor will the tools be in place to fill any gaps between the actual and most profitable desired availability. Nor can there be effective improvement and change cycles throughout the plant's life.

Sources and Forces of Change

Furthermore, where will the forces of change and improvement come from? These lead the processes to study, realign and optimize the plant's design and functioning. This question also applies to the forces to maintain the integrity of the most recent availability scheme. Thus, plant performance will progressively move away from the specified availability rather than move towards it. Nor would the forces to press for change toward the desired availability position be adequate to counteract such movement.

Short-Fall of Potential Performance

This means that the corporate, business and plant management cannot help but be substantially less effective than what is possible. This makes it further less likely that the plant will ever fill its potential for maximizing the owner's organizational profitability, productivity of assets, and other social responsibilities.

Dominance of the Drive for Proficiency

So what is management left with? First, it is probably reduced to managing for availability with stress on maintenance proficiency. In other words, making and supporting efficient and effective repairs. This is in conflict with functioning within the constraints of achieving an organizational optimum. Attempts to discover what plant availability is, what it should become, and how to get it will possibly be resisted because of the stress on proficiency. This will be manifested in very subtle and unconscious ways since maintenance operations will represent the full domain of availability management rather than one within it. The organization will limit its business to proficiency rather than most profitably delivering the currently specified availability the plant is able to produce.

Worse, these organizational dynamics will probably be very effective in their resistance. This is because management will have in place few, if any, of the necessary tools and functions to serve the need for maintaining the integrity of the availability scheme and developing changes to that scheme.

Summary

Some final comments in conclusion of this section and the book. The traditional discussion of availability engineering and management would have seen the commercial production phase as the continuing planning and deployment of the plant maintenance and support functions. That vision would have also included the evaluation and improvement of those logistics with a collateral improvement of plant economics.

However, this section has stepped out of that perspective to consider the whole picture of production system management as a set of information, decision, and management cycles. From there, the chapters defined the associated availability engineering and management cycles.

To explore such a view is new ground. It is an important one because the primary design issue for production facilities is their use in commercial production to accomplish the owner's strategies. This is a strong departure from design issues that are limited to the vision of efficiency and costs of the design, construction, and startup phases.

Therefore, the project design criteria and goals should be taken from the stated operational requirements for the plant's commercial phase. Furthermore, the value of any design process should be inspected in the context of the whole scope of the production system management and its various cycles. Under such an inspection, the value of availability engineering and management looms large.

Therefore, the message of this section is simple. All roads must lead to serving the productive life of the plant. If the right things are not done during the plant development phases or later, the many classical management processes and cycles will not fully be able to do the right things. Another view, is that all design phase deliverables should be regarded as preparation for successful commercial production management.

The last three chapters have defined the availability management processes that are served by the design phase requirements. The reader may have noticed that these processes repeatedly returned to and updated the analysis and systems created in the design phases.

It bears repeating that the book is equally applicable to existing plants. The only difference is that the phase discussed in this section, the commercial life, may be the first phase to be designed. The objective is to define the availability engineering and management processes to ultimately become part of normal corporate, operating company, and plant management. The design and deployment phases should then take place as described in earlier sections.

Bibliography

Buffa, Elwood S. and Sarin, Rakesh K. *Modern Production Operations Management.* 8th ed. John Wiley & Sons. New York. 1987.

Katz, Daniel and Kahn, Robert, L. *The Social Psychology of Organizations.* 2nd Edition. John Wiley & Sons. New York. 1978.

Tersine, Richard J. *Production Operations Management Concepts.* 2nd ed. Prentice Hall. Englewood Cliffs, N,J. 1984.

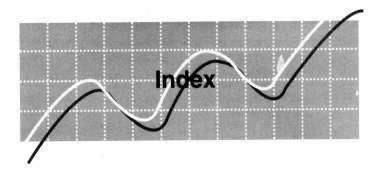

Index

A